Desert of Dreams

& Mardine

ert of

ams

enjoy this

ob Holloway

Holloway

02

BLISHING GROUP

Copyright 2002 by Bob Holloway. All rights reserved.

Printed in Canada

Publishing services by Selah Publishing Group, Arizona. The views expressed or implied in this work do not necessarily reflect those of Selah Publishing Group.

No part of this publication may be reproduced, stored in a retrieval system or transmitted in any way by any means, electronic, mechanical, photocopy, recording or otherwise, with out the prior permission of the author except as provided by USA copyright law.

ISBN: 1-58930-058-0 (soft cover)
ISBN: 1-58930-062-9 (hard cover)

Library of Congress Control Number: 2002102280

Dedication

    This book is dedicated first of all to my wife, Agnes, who spent many hours editing out my many mistakes and making suggestions to improve the quality. Then our daughter Cindy who had the dubious privilege of being raised in the environment to make a farm from the desert. I would also like to remember Ed Williamson, Ike Parker and Bob Graham who have died after our interviews and before the publishing of this book. Last, but certainly not least, my parents Earl and Ruby Holloway, who cut the ties of home and family in southern Idaho and moved into the middle of the sand dunes and sage brush of the Columbia Basin. They were the very essence of making that Desert of Dreams a reality.

Contents

Prologue	IX
Acknowledgements	XIII
Dean and Dorothy Bair	15
Marshall and Janie Field	33
LeRoy and Cleo Gossett	47
Bob and Virginia Graham	61
Dean and Russella Hagerty	67
Tub and Wanda Hansen	81
Haidi and Shigeko Hirai	93
Paul and Virginia Hirai	99
Earl and Ruby Holloway	113
Roy Hull	131
Dean and Catherine Moore	139
John Morris	155
Ken and Dorothy Murphy	157
Harley and Juanita Ottmar	165
Ike and Alice Parker	177
Larry and Eleanor Richardson	195
Mabel Thompson	209
Everett and Dee Thornton	221
Ed and Signe Williamson	231
Don and Byrdeen Worley	237
Epilogue	251
Index	253

Prologue

The Desert of Dreams

The last of the "Old West Pioneering" didn't take place in the 1800s. It was in the 1950s and 1960s, at no other place then central Washington State.

They created funding for Grand Coulee Dam under the National Industrial Recovery Act of 1933. After a delay at the beginning of World War II, it was re-authorized in 1943 under the Columbia Basin Project Act and the water arrived in 1952. The basin developed rapidly through the 1960s and 1970s with small amounts into the 1990s.

In 1955, the Grant County PUD received a 50 year authorization for Priest Rapids Dam to supply more power for the Basin and the Northwest. By 1964, the PUD was constructing Wanapum Dam. Without the power generated from these dams the potential of the entire Columbia Basin Project would not be realized. It is important to note the PUD did not build the dams with federal funds.

The project consists of approximately 553,000 acres in 1998, making it one of the largest in the western United States. It lays within an area approximately 100 miles long and 40 miles wide. The meandering Columbia River borders it on the west and south. The east and north give way to rolling dry land wheat country. It extended to the back waters of Grand Coulee Dam to the north and to northern Idaho on the east. This all makes for the most productive farm production area in the nation.

What kind of people did it take to move into this area when the irrigation started and expanded? Where did they come from? What were their backgrounds and experiences? Why did they come? How and where did they live? Did they have any hardships planting their first crops while trying to establish their own roots? Did they all prosper? Was it all work and no play?

The dreams of accomplishment were not of the settlers alone. The leaders who made it happen had a vision of the future benefits of the dam construction. These objectives had large scale economic benefits not only to the Pacific Northwest, but the whole nation. Among those goals were insuring the nation of abundant food supplies from the irrigated fields. It created flood control downstream and hydroelectric power distributed throughout the Northwest. It would enhance recreation by forming thousands of acres of new lakes. Those lakes created fish and wildlife habitat beyond what the desert had to offer. Last, but certainly not the least to mention is tax revenue to the counties, Washington State and the Internal Revenue Service.

Another unexpected benefit of Grand Coulee Dam was the electricity to run the aluminum plants during the war. This took priority over the initial purpose of irrigation until after the war ended. Many believe the dam was largely responsible for the U.S. victory of World War II.

In 1992, the gross irrigated crop value amounted to 520 million dollars. A shift to higher value per acre crops, will increase that amount in the future. This is accomplished with about two percent of the runoff at Grand Coulee Dam, with some of the water recycling two and three times and eventually returning to the river. The estimated recreation activity exceeds 22 million dollars annually. Agricultural services and food processing provide income as well as generating another 297 million in local area dollars. The irrigated land within the project accounts for ten million annually in state and local tax revenues. Taxes created by agricultural and its related factories and businesses create millions for the IRS. (Figures are from a study prepared by Pacific Northwest Project, Kennewick, WA and sponsored by the East, South and Quincy Irrigation Districts.)

Grand Coulee Dam saves millions of dollars of avoided flood damages during a severe flooding event. The back waters have created hundreds of acres of wetlands with fish and wildlife habitat. Hydropower is the best of all avenues as a source of energy. It is non-polluting, it is renewable, it doesn't deplete the water supply and the lower power costs helps all consumers. During a drought, the back waters are slowly released for power and retain flow for fish.

Despite all these benefits, many environmental groups back the idea of removing dams from the western rivers. The dam removal supposedly would help the salmon. Despite all these benefits, many environmental groups back the idea of removing dams from the western rivers. The dam removal supposedly would help the salmon. The Grant County Public Utility District owns two dams below Grand Coulee, Wanapum and Priest Rapids. (A video documentary produced bye the Grant County PUD is available upon request.) Their 1999 budget called for more than 22 million dollars for salmon enhancement. Within the last several years they have spent 150 million for this purpose. This has been done with the encouragement and demands of the environmentalists and many of the Indian tribes.

The modern farmers are no longer the pioneers of the past. Yet they have a new set of problems and changing values to face and overcome. They need to solve the problems to forge ahead and maintain a healthy and adequate food supply. The need to secure hydropower for homes and industry are critical.

This book attempts to give different perspectives of incoming settlers. It will touch on their lives before they came and some times it will say why they came. We shall try to know these people better. Perhaps we can see how the irrigation project affected the lives of those who dared to leave their homes, families and familiar surroundings to start anew in the desert of dreams.

Acknowledgements

The inspiration to begin this writing was provided by James Lafayette Holloway, a cousin from California who has completed several historical documents on the early life and times of people in Southern Idaho.

Thank you, Eleanor Richardson, Quincy for proofreading the text. I am indebted to Karen Bonaudi of Moses Lake for preparing the photos for publication.

Many hours of work were performed by Dick and Janet Schafer of Moses Lake editing and preparing the text and catching the oversights I have made.

Most of all, I want to thank all the people who shared with me. We visited while the tape recorder tried to pick up a small portion of their lives, adventures and dreams of their families past and future. They were all patient and understanding. It was a privilege to put their story into words, inadequate as they may be.

Dean and Dorothy Bair

A Story Of Two Bairs

Ask five different people, what is success? You are likely to get five different answers. Some measure it in financial terms, others may determine social status being at the top of the list. Many feel their religious beliefs are the ultimate measure. Nearly all of them will include how well they have instilled their values into their children. Values such as love, honesty, responsibility, respect for God, themselves and each other. Success doesn't come easy, but some enjoy the effort more than others.

Dean Bair was born at Richmond, in Cache Valley north of Logan, Utah, in 1915. He had four brothers, Leigh, Wayne, Myron and Francis Leon, Jr. and two sisters, Connie and Grace. His parents, Francis Leon and Laura Ethel Bair, were farmers. When Dean was quite young, his dad, everyone called him F.L., was interested in a better farming area. F.L. and a friend traveled to Twin Falls, Idaho where new farms were being developed. When they arrived, the wind was blowing huge dust clouds over the landscape; it looked miserable. In all, it didn't make a good impression. Then he heard of a farm at Shelly, Idaho, north of Pocatello that was already developed and for sale. After inspecting the Shelley farm, he bought it. Dean was three years old when the family moved north in 1918.

The farm at Shelley was rill irrigated. They raised cows and row crops, potatoes, sugar beets, grain and hay. In addition to the farm, F.L. started a coal supply yard. He sold and delivered coal to town residences. The coal supply came from Utah. A few years later, F.L. opened a fresh potato packing shed. The fresh shed was used to run their own and other farmers' spuds. There were 11 fresh potato packing warehouses in the little town of Shelley. It was the potato packing capital of Idaho at the time.

His younger sister, Connie, died when Dean was 13 years old. When he went with his father to Utah to pick out a headstone for her grave, the last thing his dad told Dean's mother before he left was, "We might as well get a headstone for my grave too." Shortly after that, F.L. went fishing in Logan Canyon with Dean's grandfather. F.L. had a cold before they left and it got worse while he was gone. He went to bed at noon, the day after returning; the next morning he was worse. A doctor said he had pneumonia and told them to take F.L. to the Logan Hospital. Their new Studebaker car quit on the way to the hospital. No one would stop for them until Dean's grandfather stood in the middle of the road and waved someone down. They put F.L. into the other car and transported him to Logan, but he never recovered.

Dean's uncle from Driggs, Idaho, took over running the warehouse until Dean's mother decided to sell the business to Tom McCarroll. Dean went to school in the winter and at the age of 15, he and his brother took responsibility of the family farm during the summers. When he graduated from Shelly High School in 1933, he worked in the sugar factory in the winter until he was 20 years old. After that he worked for Tom McCarroll driving a Ford pickup truck delivering Aberdeen Coal during the winter. If the order was over a ton, he used a larger single axle truck to deliver to coal bins normally located beneath a home. He knew the location of every coal chute in Shelley. He also drove a potato truck for Tom during harvest season.

To help with finances, Dean's mother took in school teachers for room and board. She needed help and was fortunate to hire a dependable young lady for three dollars a week, which was good money at the time. Her name was Dorothy and she was the daughter of Frank and Martha Anderson of Shelley. Dorothy worked after school and full time in the summer. Her father worked at various jobs around Shelley. She was born in 1921 and had six sisters and one brother. She was six years younger than Dean. Working for the family, she learned to love his mother before she did Dean. Dean and Dorothy were married in 1937 at the LDS Temple in Salt Lake.

Dorothy could have made it in show business with her one-liners. She said, "We were married during the depression and we've been depressed ever since." "Other than knowing I was 16 when we got married, he also knew I could make good bread, scrub floors and keep a clean house." Of course, being good natured, honest, attractive, hard working and having a delightful sense of humor may have had something to do with his attraction to her too. Getting married at 16 years of age didn't allow her to finish high school, but like she says, "I'm smarter than the average Bair." On the serious side, she thought the best thing that ever happened to her was marrying Dean.

Dorothy's father also had a sense of humor. When Dean asked his permission to marry her, he said, "Do you think you can get along with her, none of us can." Dorothy is great to get along with and Dean was fortunate to find such a terrific lady right on his own doorstep, and they have been getting along for 63 years. Of course, Frank Anderson already knew that.

They continued farming and working in Shelley until 1942, when they moved 100 miles west to the Lost River country. They bought some ground between Arco and MacKay, Idaho. The elevation was 6000 feet which was good mostly for cattle and wheat. Dean felt the isolation and elevation could be beneficial for seed potatoes as Dorothy's brother-in-law had raised some fine seed nearby. At first they were going to rent ground, but when Dean talked to the owner, he said, "I'll sell it to you cheaper than you can rent it." The country had just been through the Big Depression and things were still tough economically. Dean told him they didn't have any money. The man said, "Do you have $50?" Dean said, "yes," he had that much. The man said to give him $50 down and $150 a year and I'll sell it to you. They bought the quarter section of irrigated ground and that much more of dry land.

There was an old house on the place. A dirt basement had been dug beneath the house for grain storage with an opening to the outside. It was completely infested with mice. Dean filled all the holes in the wall with plaster. The next day they were open again. The next time he mixed the plaster with broken glass, the holes stayed sealed. The outside walls were covered with unpainted six and eight inch boards that were warped. From the outside, they could see the plaster on the inner walls.

They moved February 1 of 1942. Dean drove their Oliver 60 pulling two wagons, one with their machinery and the other with 96 sacks of seed potatoes with triple layers of tarp over them to keep them from freezing. It was a cold and windy day. The Oliver was completely open with no protection. Dean put on all the warm clothes he could, knowing it would be a cold ride. He started from his mother's farm in Shelley at four a.m. on a 100-mile jaunt. He finally got to Midway, which was a little beer joint halfway across the desert. He had never been that cold in his life, he was almost frozen when he stopped there at 2:00 p.m. and went in to warm up. The owner of the bar could see how cold Dean was and offered him a shot of whiskey to help warm him. Dean, being an upstanding member of the LDS church, thanked him but said he had never drunk alcohol in his life. The man reached under the bar and handed him a can and said, "Drink this anyway."

Dean bought some cookies, stood by the stove eating cookies and drinking his once in a life time can-of-beer, trying to warm himself before continuing down the road. When he walked out of the door, he had a tough time walking straight. For the next several miles down the highway there were two center lines where before there had been one.

The two wagons were a load for the Oliver 60. That became more evident on two different hills Dean had to go over. He got part way up the hill to find the tractor wheels slipping and his progress stopped. He had to back down and try again, the second time he made it to the top. The difference had been the paint on the yellow center line was slicker than the regular road surface. The first time he had one tire on the line causing it to spin out.

As the afternoon passed, Dean was again almost frozen. He tried to ride in such a way as to catch the heat from the exhaust pipe without much success. He actually got off the tractor and ran along the side, between the front and rear wheels while reaching over and steering with one hand. At least he was so cold he didn't get sleepy. It was midnight when he arrived at a neighbor's farm. The neighbor got out of bed and helped him get his potato seed backed into potato storage. Then Dean went home and tried to warm up without much success. The only thing he could do was go to bed with his teeth still chattering and eventually got back to normal. He didn't even get a sniffle from the cold, 100-mile trip, and he only froze four sacks of seed in the process.

They spent the first four years in the house without electricity or inside plumbing. The well had a Galloway gasoline motor to pump the water to the surface, then carried by bucket into the house. A few days after moving in, the temperature dropped to 30 degrees below zero. To keep warm at night, they confined their living quarters to the small bedroom containing their bed, Ross's crib and a coal stove. Dean could reach from the bed to throw a chunk of coal in the stove once in a while throughout the night.

During the first year they stuccoed the outside walls, put new shingles on the roof, put windows in the basement and installed a new coal furnace. The next year they installed storm windows in the house and built a new cow barn. With a lot of work, they turned it into a quite good farmstead. They were milking ten cows by hand at first, but finally got a milking machine with a gasoline motor to run it.

Dean and his brother F.L., cut and peeled logs from the hills, hauled them down and built two new potato storages, one for each of them. They were partially underground. The roofs had mesh wire over the rafters to hold wheat straw and covered with a dirt surface. It stayed warm in the winter and cool in the summer.

The ground had rocky soil. They remained there five years before moving to Leslie where the soil was much better. But it was still a high and cold climate. Dorothy said of the country, "It had two seasons, July and winter." One year on the 26th of June, their potato rows were about closed when the morning temperature dropped to 28 degrees. It froze 100 acres of seed potatoes to the ground. They recovered only because the seed had enough vigor left to sprout from the seed piece and grow again. The yield that fall was only two and a half ton per acre, however they lucked out because the price was five dollars per 100 lbs. which was very high at the time. They decided they wanted out of that cold country. They put the farm up for sale and never got one offer in three years.

Each fall, Dean, and friends, went elk-hunting in the Salmon River country and across the line into Montana. They hired a guide that furnished horses and packing gear for sixty dollars apiece. The fall of 1951, Frans Yorgenson, an old friend of Dean's, originally from Shelley, farming in the Lost River area, wanted to join them in the hunt. Newel Anderson also joined them. During the hunt Newell and Frans told Dean about their visit to the Rosa Irrigation Project near Yakima, Washington, where they helped Dan Cox harvest his potatoes. The yield had been twenty tons per acre. They related to him how big and nice the tubers were and within just a few feet, the picking basket was full. They said it was wonderful up there, the spuds grew bigger, the hens laid more eggs and the women had more babies. It was the land of milk and honey. The whole hunting trip was consumed of the talk about how great Washington was and their words fell on very receptive ears. They said the Columbia Basin Project was opening the next spring. It was the first Dean had heard about the new irrigation project in Washington State. After listening to Frans and Newell, Dean was hooked.

Their farm had a nice house on it. The neighbor was short of water on his farm and glad to get the chance to buy the Bair place. They made the sale and made a quick trip to Prossor, Warden and

Quincy, Washington, the day after Christmas with Newell, and Cody Anderson. They were impressed, only a skiff of snow was around and green vegetation was evident. Compared to the home place it looked terrific. Dean and Newell bought a section of ground at Warden, but it wouldn't have water for three years. At Ephrata, Dean became acquainted with Billy Clapp who took a liking to him. Clapp's daughter was married to Ed Nasburg who worked for the Bureau of Reclamation. The Nasburgs owned three units near Ephrata which would have water the next spring. Clapp, who was overseeing the property, rented the ground to Dean.

By midwinter, Dean, Dorothy and the four kids were prepared to migrate westward. Dean rented a railroad boxcar for $700. He built an upper deck and loaded their furniture on the upper deck and the farm equipment below. Another boxcar cost $400 and he loaded it with potato seed. It was just like ten years earlier, only this time he was having the railroad do the driving.

The family came by auto. The snow was three feet deep as they left in the middle of a snow storm. A snow plow came out of MacKay to get them out to the road. They were the last car to make it over the highway past Craters of the Moon heading west. The first night they stayed with Dorothy's folks in Jerome, Idaho. The second day they made it as far as Baker, Oregon. During the afternoon, they passed the train carrying their furniture and equipment in southeast Oregon. They arrived in Ephrata the following day, February 17, 1952, on their son Keith's birthday. Their belongings didn't arrive for another week. While waiting for the furniture to arrive, they stayed in the Hi U Motel at night and took the kids to the empty rental house on 64 G Street SE during the day.

Dean said he would never forget the feeling he had about two weeks later when he drove to work at the farm. The Bureau of Reclamation made ditches to the high point of each farm unit. He pulled up to that spot and parked. All he could see was sagebrush and cheatgrass. He wondered to himself, "What kind of fool am I for leaving a settled farming area in Idaho and coming to this?" He then reasoned, there was no turning back. His pioneering blood took over and he decided to get to work and make the best of it.

Larry Armstrong, a teenager from Idaho arrived. He said there were numerous storms after Bairs left. After every snow, they got wind, piling the snow deeper. The schools had been closed for two weeks. Larry had come to Washington to work. After helping two weeks, Larry returned home and it was another two weeks before school could open again.

Dean cleared and leveled the first 40 acres on the Nasburg land. After preparation, he started to plant peas. He had Larry plowing on the field across the road. It was a beautiful morning, but later he noticed dust blowing west toward Quincy. By 11:00 he couldn't see the mark of his last pass through the field with the drill. The wind was then blowing like crazy. It was impossible to see anything. He pulled the pin on the drill, but he couldn't even see where the sun was in the sky. When he tried to leave the field and return to the shed, he didn't know which direction to go. By getting down on his hands and knees he finally found a track, enough to find he was going the wrong direction. After getting out of the field, there wasn't a bit of dust over the ground that hadn't been worked. When he and Larry returned that evening, the wind was still blowing. The dirt on the ground that had been worked was being carried downwind. As soon as they walked out in the field, they couldn't even see each other.

The next morning the wind had quit and they checked the field. The peas had been planted two inches deep the morning before, but they were all on top of the ground. The soil had blown away, but the peas never moved. U.S. Air Force planes reported dust clouds the day before up to 12,000 feet. At least Dean knew where his two inches of top soil went. They used the disc to cover the pea seed and corrugated the field. A week later another wind hit and flattened the corrugates. Dean corrugated again, then the first water arrived and they were able to hold the loam soil from further blows. It was among the first water deliveries in the Columbia Basin Irrigation Project. The odd thing about it was, Dean said, the field produced the best crop of peas he ever had, in more than 40 years of farming. He never had a crop to match that first one. The peas were sold to Brotherton Seed Company in Ellensburg. After the first year, the landlord built a house on the farm. The first year had been great compared to Idaho.

They were able to contract sugar beet acreage from Amalgamated Sugar Company the second year, 1953. Roy Hull of Quincy was the company representative. Raising a good crop was not easy in the early years. It was the first crop planted in the spring and the last harvested in the fall, sometimes under cold, wet and miserable conditions. Wind caused problems until they were large enough to hold their own. Weeding was a constant battle. Getting a good field crew for weeding and thinning always took a lot of supervision. But they knew how to raise beets and did a good job. It was always a solid, dependable source of their farm income.

About 1954, the highway between Quincy and Ephrata was in the process of being widened. Large road equipment had the roadway torn up when another wind storm hit. The farm, and their house were next to the highway. The wind came at an angle that dust filtered into the house so bad they were afraid their baby would get consumption. They were able to drive down the canal bank road to get around the construction and went to town to stay with friends. After dark that evening, when the wind died down, they returned home. The house had weather stripping around the doors and windows and was pretty weather tight. The living room carpet had flower patterns, but they couldn't even tell what color it was from the layer of dirt covering everything. It looked like a dirt floor. Their newly established lawn had piles of soil over it like snow drifts. Dean used the tractor and scraper to uncover it.

While the construction crew was working on the highway near the house, their son, Keith, learned to be a business man. He toted out water to the workers who gave him nickels and dimes for his efforts. He was about nine years old and thought he had found his gold mine.

One spring, Jack English from Ephrata helped plant potatoes east of the house. West of the house, Dean had a field leveled and ready for his son Lyle to start corrugating the next day when he was out of school for the weekend. Jack and Dean had only a few rows of potatoes planted that morning when he noticed the sky to the north turning into a yellowish haze. Dean stopped planting and told Jack, "The wind is going to blow, I've got to start the corrugator to hold that leveled field or its going to be trouble."

By the time he started the corrugator, the wind arrived. The dust from the worked up field was blowing clouds of dust across the highway. Before he finished a couple rounds there was an accident from lack of visibility. Dean knew of the wreck when a man with blood on his face from hitting his steering wheel walked up to the tractor and cussed him out for raising so much dust with the corrugator. His nose had been bloodied and it was dripping onto his white shirt. He wanted Dean to stop stirring up the dirt. It was making things worse. Dean walked over the corrugated ground with him to show how it had stopped the blow compared to the undone portion that was so bad. The man understood, apologized and told him to keep going, and left.

It wasn't long before more accidents took place. A National Guard representative appeared out of the dust. He said a Washington State Patrolman sent him to tell Dean to quit stirring up the dust. Again, Dean showed him the difference where he was reducing the blowing with the corrugations and the man apologized and left. The next one to appear was the patrolman himself. He chewed Dean out and said, "You've just got to stop this." After letting him see both areas of the field, the patrolman also apologized and asked him to finish.

The following day a patrolman knocked on the door asking if this was where the dust was so bad yesterday. He was trying to locate the exact area of the accidents for the records. Seven cars had piled up entering the blowing dust. They hadn't been able to see the car in front that was stopped, and kept hitting the one stopped in front of them. Thousands of dollars of damage had occurred. Needless to say, Dean tried his best to prevent it from happening again. After some of the windstorms, road graders cleared the roads of sand accumulations.

Their first crops in 1952, other than peas, were red clover seed and potatoes. They had some good crops after learning how to manage the ditches and set syphon tubes. The potatoes didn't do well the first year, but the potatoes following two years of clover were fantastic. They averaged 70% more than 10 oz. for size. Boyton Dodge ran the crop through his fresh packing shed in Winchester. He told Dean they were the biggest Russet Burbank potatoes he had ever seen. They dug four spud rows into one, then the tubers were hand picked into wire baskets. Two baskets were emptied into burlap sacks. The sacks were so close together that another sack wouldn't fit in between. The field was next to the highway and cars would stop and inquire what type of crop was being harvested.

The 1955 potato crop was another excellent yield. The landlord told Dean not to worry about frost until after Thanksgiving, so he put off digging spuds for a couple of weeks. To earn some extra income he did custom work harvesting his neighbor's beets. When Dean started on his own potatoes again, he hired 24 Mexicans to pick by hand, and two mechanical harvesters loading bulk into trucks. The weather was still good and harvest was going rapidly.

It was the 10th of November. When they finished that day there were still 180 sacks left in the field the last truck didn't have room to pick up. The weather had been warm and Dean wasn't concerned about leaving them out overnight. A cold north wind came up during the night. He got up to check the temperature occasionally during the night. When he got up at 4:00 in the morning to get ready for work, there were two inches of snow on the ground. The temperature had dropped to freezing level. Dean rushed to his brother's (F.L.) house, got him up and they loaded up the spuds that had been left out in the field. F.L. took the spuds to town while Dean started the beet harvester, hoping it would warm enough by noon to start potato harvest again. It didn't. By 4:00 pm, it was so cold, the Marbeet digger that spiked the beets and brought them up to the topping unit was losing most of the beets. They were frozen to the point where they wouldn't spike. There weren't any cabs or heaters on the tractors then and Dean was freezing. F.L. came out and said, " I wouldn't sit out here and freeze like that for anything." Dean replied, "Brother, this might be the last beets we harvest this year." And it was! That night it went to 10 degrees below zero. The next night it was 20 below and it didn't warm up. Harvest was over.

He hoped to harvest the beets in the spring, but they had turned to mush. Dean sold them to Willard Lange to pasture his sheep. They also lost 40 acres of potatoes. Only 40 tons of beets were harvested and they lost all of the clover seed crop. That was the year, Dean said, that we went broke and wouldn't admit it. The family motto for the next few years was "If it costs money, forget it." It didn't help, but they weren't alone.

To raise clover seed, the plant has to be pollinated by bees. The bees were brought in by the truck load from other areas, sometimes as far away as California. The hives were unloaded near the field and after the season was over, loaded up to be transported some where else. One evening Dean agreed to help the beekeeper move them from another location to the clover field. When they were ready, they put on the bee headgear with a net that came down around the face and shoulders to prevent bee entry and avoid stings. As they were working, Dean's hat kept slipping down, he kept pushing it back in place. Finally the net tore in the back and allowed bees to enter. Before the night was over, he had been stung 17 times on the neck and face. Some bees had crawled up his pant leg and stung him on the legs. At least the beekeeper occasionally brought over a five gallon pail of honey for them which they shared with friends and family.

Dean related the story of two young brothers who sold everything they had in Utah and came to Washington to make their fortunes. They bought sandy ground along the highway south of Quincy. They ordered water for a Monday morning. When George Scott, the ditch rider arrived to turn on the water, he saw one of the brothers sitting next to the head gate. He was dejected almost to tears. The farmer told him, "Don't bother turning on the water, we planted three times and been blown out three times. The ditches are full of sand and we're through. We can't borrow any more money. I guess we'll have to go back to Utah and see if we can find a job to pay off our debts." The ditch rider said, "It looked like he had been bawling and I felt like sitting down and bawling with him." The Bureau of Reclamation had predicted that three of the first four farmers on each unit would go broke on the development of the Columbia Basin.

Dean and Dorothy had a chance to buy 80 acres at the appraised price of $7,800 from Harry McMullen in Block 89, southwest of Ephrata. They didn't have the money for the purchase. Dean asked Billy Clapp for a loan. Clapp said, "Do you think I'm crazy? Why would I loan you money for your own farm and lose a good renter?" Dean told him he would move eventually anyway, you might as well help me. Clapp sat down and wrote him a check. Dean asked if he wanted a signed contract. Clapp said, "I'm honest and I think you are, that's good enough for me."

They farmed both places until building their own home on Road F Northwest and moving there in 1959. More ground was bought from the Bureau of Reclamation, Verne Mathews from Ephrata and an owner from Seattle, Mr. Reninger, bringing the total to 400 acres. The Mathews ground had been farmed by Russ Patton before they bought it and was the only ground they didn't have to clear themselves. When called on the phone, Reninger refused to sell his eight acres which was landlocked in the middle. He was going to rent it to his son. Dean and Dorothy drove to Seattle to see him. They weren't able to talk to him, but they visited with his secretary for 30 minutes. The following day, Reninger called them on the phone and said, " I don't know what you told my secretary, but she convinced me I should sell that ground to you. Send me a check for the going price and its yours."

While rotovating in the sagebrush one day, he ran into an old sheep corral that had wire fences. The wire was buried in the sand and the implement wrapped it around and around so thoroughly that Dean stopped and walked to the shop. When he returned he couldn't find the tractor in the tall sagebrush. He finally had to follow the tractor tracks where he first entered the brush to find it. Over the years, they've cleared brush from nine quarter sections.

Dorothy said they arrived in Washington with four kids and now there are 85 of us and counting. That in itself should provide proof of the Basin's productive capacity. Dean and Dorothy had five boys and one girl. They have 34 grandchildren and 30 great-grandchildren. There are nine families of Bairs living in the area. Arlene and Jay were born in Washington. Arlene is married to Eric Johnson, a banker, and living in Dayton, Washington. Glen is farming and is a talented writer and performer of cowboy poetry.

Dorothy said, "When most kids get married, they leave home. When ours got married, they came home." After finishing college, Lyle and Glen wanted to farm. Dean told them, "I want you to be certain it's what you want to do. You work for me for two years as hired help, if you still want to farm, we'll help you get started." Ross worked for Clark-Jennings and Lamb-Weston before raising seed spuds in Montana and eventually came back to the Basin to join Lyle in farming at Stratford.

Keith worked for Lad Irrigation in Moses Lake and as a Washington State potato inspector. He told Dean he felt like he was left out, he thought he would like to farm too. They made the same deal, work two years as hired help and then take over one of the units. After farming several years, he rented his ground to his brothers and took a job driving truck for Zip Trucking in Moses Lake. When Jay returned from his mission for the church, he also wanted to farm and his folks started him out with a unit. He went on to farm and also works for Dave Canfield, who owns an insurance company in Ephrata. Jay also is the pilot of the company's two engine aircraft. Jay was one of the founders of Grant National Bank.

The sugar beet plant in Moses Lake restarted in 1998 and closed before the season in 2001. It was a gallant effort by the growers to revive the industry. After individual farmers funded the start-up and didn't receive a large part of the crop returns for three years, it failed financially. The plant had mechanical problems the first two years. The third year it ran very well. However the extremely large debt load along with a worldwide surplus of sugar was too much. Many growers and agribusiness people were hurt financially, including the

Bair brothers. The spring of 2001 had several farm auctions as a direct result.

When the boys wanted to start farming on their own, Dean told them, "I'll help you get started, but if I ever hear one word of discontent among you, we'll sell the whole thing." All of the boys have been involved in farming, true to their upbringing. They are a credit to the example Dean and Dorothy have set for them.

Dean served on the Quincy Irrigation District Board and has been a long standing member of the Farm Bureau. Dean served on the Grant County School Reorganization Board and was the first Bishop of the LDS Church in Ephrata. In 1982 and 1983, he and Dorothy served on a mission for the church, seven months in New Mexico helping with an Indian tribe in farming and 11 months in Missouri. Otherwise they have kept their work confined to the children, the church and the farm. In between, Dean took up flying and earned his single engine, private pilot rating. Dean and Dorothy turned the farming operation over to the boys in 1977 ending almost a half century of active farming.

Dean and Dorothy are successful. Not because they raised some great agricultural crops through the years. Not because they endured the hardships of developing new productive farms under difficult circumstances. Not because they were faithful to their beliefs, or followed the Golden Rule. Not just because they helped their children get started on their own life's dream. It was because in doing all the above, they were a living example that taught through actions and words. They succeeded in raising a human crop of loving, responsible, respectful, hard working, talented and compatible children of which any parent would be proud.

Jay, Keith, Ross
Lyle, Dean, Dorothy, Arlene, Glen

Harvesting the first Washington
potato crop - 1952

The four big Bairs in the Bair's big beets.
July 4, 1966

60th Anniversary

The following poem was printed in the Grant County Journal newspaper in the past.
Dorothy wanted to share it. We couldn't find the authors name.

Grant County has a history as some of you may know,
But many never stop to think what made this county grow,
An "old timer" once told me of the days that used to be,
And now I'll try to tell it the way 'twas told to me.
Indians lived and hunted throughout this countryside.
Cattle and wild horses too roamed here far and wide.
The River Columbia, from Canada its source,
Flowed lonesome in its valley, unhindered in its course.
Ephrata was platted in nineteen and one
And with that chore completed, its history had begun.
The town was always destined to live up to its name —
An oasis in the desert, like the one of Bible fame
The railroad's boxcar depot insured the town would stay.

32 / The Desert Of *Dreams*

A very small beginning for what we have today.
Stores, saloons, and blacksmith shops followed, swiftly then
As well as schools and churches, to serve the needs of man,
The county then was Douglas, so big from side to side
The citizens decided 'twould be better to divide.
Grant County is the part we got, our town the county seat
Though this was still uncertain, 'till a vote saw Adrian beat.
The bank was robbed, the town caught fire, a new courthouse was planned.
All of these, in nineteen-nine made history in this land.
Cattle was the mainstay then, and in the highlands wheat.
Though many of the farmers by lack of rain were beat.
Far-sighted dreamers then could see no future for the land
Unless the drought could be relieved, so Coulee Dam was planned.
Many year abuilding, but came the awaited day
When gates were finally opened, and water flowed its way
To land where early settlers, many years way back
Had left their farms and homesteads solely from its lack.
Although it may seem commonplace to folks like you and me,
People travel 'round the world this miracle to see.
But we must all remember this work is never done,
And for my generation, it hardly has begun.
I hope that we who follow can accomplish as much or more
As the pioneers and dreamers who passed this way before.

– authur unknown

Marshall and Janie Field

Watch Your Coffee Cup

The B-24 was above 20,000 feet over France when 8 mm flak from enemy fire scored a hit. The large craft shuddered and smoke started billowing from number three engine. The young American pilot fought the controls which weren't responding. A quick check found there were no serious injuries to the crew. However extensive damage had occurred to the aircraft which was now losing altitude and dropping down out of the formation of bombers. Marshall had a sinking feeling in the pit of his stomach, but he was in charge and trained for just this eventuality.

He gave the orders to bail out. Marshall thought he was the last one to parachute from the doomed aircraft. Unknown to him, his engineer hadn't jumped yet. By the time he did, the altitude didn't give the chute a chance to open and he was killed upon impact. Marshall and another crew member who jumped with him landed in a grain field. Still breathless from the parachute drop, they quickly buried their parachutes so no one could find where they landed. They were fortunately separated from the rest of the crew who were captured by the German troops.

The plane was shot down during daylight. Marshall and his friend hid in the field from the German patrols looking for them until it was dark. Morning arrived after a long sleepless night. They discarded their military clothing and started walking. They came across three Frenchmen working in a nearby field. They couldn't understand each other and used sign language to communicate. One of the farmers had them hide there until he came back at midnight. He took them home to hide them from the German search parties. The farmer and his wife sent their young daughter to her grandparents so she wouldn't accidentally give them away. For 31 days the family fed and kept them in the attic of their home. German troops conducted searches for the downed fliers. They searched the home more than once, but without checking the attic.

General Patton's army coming from Normandy, liberated France and the two fliers, ending the stay with the courageous French farm family who risked their own lives to help the two U.S. airmen. They returned to their base in London where they found the rest of the crew had been released unharmed from the concentration camp. After they arrived. Marshall called his dad in Colorado. It was his dad's birthday, what a great gift, his parents had been informed he was missing in action. Marshall's grateful mother corresponded with the French family for many years. All of the return letters she received were written in French. An interpreter was required for both parties.

The Marshall Field story actually begins almost 22 years earlier on January 16, 1922 in the small town of Lewis, 12 miles from Cortez, Colorado. His father and mother were Arthur and Addie Field. Arthur was a building contractor, building homes, churches and schools. Marshall had a horse and loved to ride. After two years of high school at Garret Ridge, Colorado he went to Las Cruces, New Mexico where he stayed with an uncle. He played basketball during his last two years of school. During that time he worked at his uncle's restaurant after school and on weekends to pay his room and board.

Upon graduation from high school in 1940, he went to San Diego, California and worked at Consolidated Aircraft producing B-24 bombers. He was at the factory until the Japanese attack on Pearl

Harbor. When he heard of the attack, he immediately tried to enlist in the Air Force. They had too many men trying to enlist from California. They wouldn't accept him but he was told he could get in quicker in his home state of Colorado. He left for Pueblo, Colorado. He was accepted in the U.S. Army Air Force.

His basic training took place at San Antonio, Texas. From there they transferred him to Muskogee, then Coffeeville, Kansas. After graduating from flight training at Altis, Oklahoma, his next stop was Fort Worth, Texas, to checkout in the same type of aircraft he helped make in California. He was assigned to pilot a B-24 bomber and sent to London.

Marshall flew with the 8th Air Force. Their base was in England where the bombing runs were to Germany and France. On the 7th mission at 7:00 o'clock in the evening, their plane was hit by flak. The plane was badly damaged, but Marshall nursed it back to base and landed. Instead of being praised for saving the plane, they reprimanded him or not baling out and ditching the aircraft. The 8th mission was his last one when they baled out over France.

It was December of 1944 when Marshall returned to the states from England. They gave him a furlough and he headed for Colorado and Christmas at home. He attended a local dance during his visit, that's where he met a tall, slim and very attractive young nurse. That evening started a hot romance. A series of dates and letters followed, ending three months later in a March 14, 1945 wedding in Phoenix, Arizona.

The attractive young lady was Janie Ertel from Cortez. She was born February 8, 1922. Dr. Trotter helped with the delivery in the family home behind the funeral parlor. The father was disappointed because he wanted a boy. A few minutes later the doctor called out, "Walt, get in here, here is another one." This time, it was a boy. The Mancos, Colorado, undertaker and his wife were both happy, with twins, Jack and Janie.

Walter and Ida Ertel eventually ended up with three girls, Virginia, Betty, and Janie and two boys, Jack and Walter. Janie attended school in Cortez, Colorado. She was the tallest girl in school and excelled in basketball. The powers to be at the time, decided basketball was too strenuous for girls. They only allowed them to play intramural games. Being very athletic, she also enjoyed competing in hop-skip and jump, running and broad-jumping. She played the position of catcher on the softball team.

After graduating from high school in 1940, she studied four years in Colorado Springs and graduated as a registered nurse. She moved back to Cortez and worked in a 10 bed hospital for a year. One night in Cortez she attended a dance at the Legion Hall. A tall, lanky serviceman was there, who changed the rest of her life. He was good-natured, fun loving and, for good measure, taller than she was. His name was Marshall and was home on leave for Christmas.

Shortly after they were married, Marshall was assigned to Arlington, California. He worked as a flight instructor for the Air Force and Janie worked in the Arlington County Hospital. The war came to an end. Marshall wanted to keep flying, but there were too many pilots. He received his discharge in August of 1945 at Pueblo, Colorado.

They moved to Cortez where Janie worked at the hospital for several months. Marshall's first job was working on a farm for the McAfees. They had sheep and Marshall hated sheep and being up night after night during lambing season. Their first child, Marsha, was born there in April of 1946. They moved to Lewis near Marshall's parents in the summer of 1948. He worked there for another farmer, Charlie McAfee at Yellow Jacket.

Charlie was a pinto bean producer and had a warehouse that handled wheat and beans. Marshall had lifelong back problems that started at the warehouse lifting wheat and bean bags. Charlie was not only his boss, he was his friend. Working for Charlie spurred Marshall's interest in farming. They were still living in Yellow Jacket on October 4, 1948 when Eddie was born. They were doing well enough to buy a new Nash car of which they were quite fond.

They had the chance to rent a farm west of Lewis called the Lord's Place. They moved there in 1951. Kelly was born in 1952 while they were farming there. Irrigated alfalfa was one of their crops, but water was scarce and sometimes by second cutting they were out of water for the year. Marshall was interested in cattle. They made a trip to California and brought home 165 Holstein calves in a pickup with a double-deck on it. The calves cost very little to buy and the first trip most of them survived. The second trip, they lost a few head to disease. On the third trip, even more of them died, so they quit. When the first calves were old enough to milk, they separated the milk and sold cream to a place in Salt Lake City. They fed the skim milk to pigs. Janie also sold cream and eggs locally for their living expenses. Things were tight financially.

Marshall loved horses and working for Charlie gave him the opportunity to enjoy them. They had horses of their own on the Lord's Place. Marsha and Eddie even rode three miles to school on nice days. Marshall continued riding until after his second back surgery years later when the doctor told him to stay off of them.

They had a small problem on the Lord's Place. It wasn't quite as heavenly as the name indicated. A family of skunks took up residence under the house. Mating season was during the winter. The family could hear thumping under the floors and soon afterward, a terrible odor would permeate the house. Janie tried burning sugar on the stove and other unsuccessful ways to rid the house of the terrible stench. She quit going to church because her clothes reeked with the smell. Marshall finally caught the skunks out in the field one day and shot part of them and eventually got them out from under the home. The last one he shot happened to be a mother. A few nights later they heard some noise under the bedroom window during the night. They discovered six baby skunks. They caught them and put them in a cage. Janie heard over the radio, they could sell skunks for three dollars a piece. At least they got $18 for all their troubles.

The next little problem in that house was bedbugs. It was not an uncommon occurrence during those times. They thought the itching they were experiencing were mosquito bites until a friend identified one that crawled out of Janie's purse one day. They had to tear the wall paper off and fumigate the house three times to clear them out.

It was 1953 before they had indoor plumbing. Water before that had to be carried in from the well.

John Clay was a friend of the Fields. He was the source of the calves brought from California. John came to hunt deer and elk in Colorado and visited with Marshall and Janie. John talked to Marshall about the development going on in the Columbia Basin out in Washington. The new land had ample water and a long growing season. Those particular qualities were of great interest to Marshall. It was a real struggle to make ends meet where they were. The house they lived in was old. They couldn't see any monetary progress for all their hard work. The water ran out in July which ended their growing season. The dream of owning a place of their own with ample water and at a cheap price was compelling. Marshall said, "Let's go to Washington and see about it."

They made a trip to the Basin in 1957. Marshall collected soil samples to analyze for fertilizer, soil type and organic matter. He liked what he saw from the start. Washington was the place to go. "I didn't like it a bit," said Janie. To add to the discomfort, she had to have a tooth pulled when they arrived at their first stop which was Ephrata. Shortly after, they looked at Quincy, which was all it took, Quincy didn't amount to much. It was spring, they left snow in Colorado and they were impressed with the warm spring weather in the basin. They decided to make the move. They wrapped up their business in Colorado that summer and moved to Washington between Christmas and New Years day of 1957.

Marshall drove a truck loaded with belongings including a deep freeze filled with meat. They filled an old bathtub with canned food cushioned with grain and bed mattresses as packing. It was cold and the meat stayed frozen. Janie and the kids brought the Ford car. The first night they stayed in Salt Lake City. The second night they spent in Ontario, Oregon. They didn't have to plug in the freezer that night as the temperature was below zero F. They arrived the evening of December 31st, New Years Eve. The first night they stayed with Opal and Lloyd Dietz south of George. The next stop was a rented house at Murphy's corner.

John Clay came up from California. John had bought a tract of land and he stayed with the Fields while he and Marshall built a house on it. Janie took part in the construction by painting and working on the hardwood floors that they installed. John had to leave for home in California before they completed the house, but it was nearly done. The family continued on to finish the yard, plant trees, and put in a garden. When the house was complete they moved in on March 14, 1958.

Marshall and Janie bought 160 acres next to the Clay property from people who developed it, but left because of the wind and hardships. The house on it wasn't completed yet. Robbie was born September 29, 1958. Meanwhile John's wife, Bev, decided to move to the Basin from California. When the Clays moved to George, the Fields moved into the unfinished house on their own farm in October. They added onto the house in 1965, including a fire place, the outside entrance covered and the basement finished, which was started when they first moved in.

Marshall borrowed money and bought cattle the first year. The cattle were kept in the corrals on the Clay property. They couldn't pay the bank loan that winter, so the bank forced Marshall to sell them. It wasn't long before he started getting some Angus and Herford beef cattle after they were on their own farm. They fed cattle for a cattleman from Yakima and Gert Seivers from Moses Lake to supplement their income. It was also a market for part of their own hay.

Janie said, "When we moved to the Basin, I absolutely hated this area. I had such a big family at home and I was home sick." She left her seven brothers and sisters in Colorado and missed them. The basin summers were hot, and the wind storms filled her house with sand through the smallest opening. The dust clouds blotted out the sun. Sometimes the wind blew for three and four days and nights without stopping. One of those spring nights, they shifted the bed around in the bedroom to keep the sand from piling up on it from the newly planted pea field dust blowing through the window sills. Janie finally insisted on storm windows for the house which cut down on the infiltration of dust. The air conditioning didn't come until the summer of 1999.

Robbie was two months old when he had croup so bad one night that Dr. Stansfield told them to take him to the Soap Lake Hospital. They didn't have enough gas in the car to get to Soap Lake. They stopped at Jack Tobin's place and he gave them the gas to get to the hospital. Robbie was in a croup tent for four days and very close to death. He developed pneumonia which turned into double pneumonia. Janie sat with him at the hospital doing what she could. One morning she left for a short time, upon returning she found they had removed the tent and Robbie was gasping for breath. Janie was almost crazy and started yelling at the nurses, saying, "Who took the croup tent off?" They replied that the doctor had ordered it off, although he hadn't been there for two days. Janie made them put it back after they suctioned out his lungs. She than called the doctor and told him, "The baby is dying, do something." The doctor prescribed antibiotics. Another nurse from Quincy, Jeanne Weber, had some patients at the hospital and started helping Janie watch Robbie through the critical period. Jeanne and Janie have been dear friends from that time on.

They attended the Presbyterian Church in Quincy where Reverend Larry Roumph was the pastor. He came to the farm often just to visit, not to preach to them. Janie and Marshall both sat and talked about the land and crops, he was well informed on the procedures and farming activities in the area. He visited off and on at the hospital when Robbie was going through his ordeal.

The farming operation became diversified. They raised cattle, alfalfa, dry peas, wheat, field corn and briefly tried chickens, pigs, asparagus, beans and potatoes. The beans rotted in the windrow before harvest from one rain after another, keeping the vines wet too long. Their first potato crop almost broke them as the price was five dollars a ton when they sold in Moses Lake. The pigs got out of their pen and rooted up all of Janie's strawberry plants, so they had a short stay. The chickens were part of a government test which the agents didn't complete. They ended up in the stew pot, the chickens, that is.

In 1963, Robbie was in the first grade. Janie went to work at the Quincy hospital to help with finances on the farm. She worked at the hospital for 28 years. Janie at one time or another over the years, has helped care for many Quincy Valley residents. She was well known in the Quincy Valley. When she started, they assigned her night duty. After a few months, while raising the kids, and other home chores, she was not getting enough sleep. She didn't know whether she was coming or going. She couldn't take the night hours any longer. Marshall called the hospital and told them she would quit work if they didn't take her off nights, and they did.

Marsha, the oldest child now lives in Durango, Colorado, running her own business. Eddie had cattle in his blood and is the manager of Schaake's Feedlot south of Quincy. He has worked for them for 14 years. He still has an interest in the home place and helps with buying and selling. Kelly spends his time doing the farm bookwork, doctoring sick cattle and riding the pens. Robbie does the mechanical work and a little of everything else around the farm and feedlot.

Marshall loved working with the cattle. He was in poor health for many years, including a bad back that started while at the bean processing plant in Colorado many years before. Three back surgeries never corrected the problem. His back wasn't the only trouble. They operated on his shoulder three times and in the summer of 1999 he had a hip replacement. He never let his physical condition get him down, he worked hard and sometimes with considerable pain and loved every minute of it. Shortly before he passed away in 1999, he was still out driving feed truck and loading cattle. After the hip operation he was on crutches. He still drove feed truck. He fixed a horseshoe hook on the side of the truck to hang the crutches from and went about doing what he had always done.

Martha's Inn is a café in George along Interstate 90. Travelers stop for coffee or a meal. The local farmers and town people take a break and stop in for socializing. Marshall was a frequent occupant of the large table near the front door where the locals gathered. When

Marshall was present, no one dared leave the table for the restroom or a phone call. Whenever an unwary person left for a few minutes, they would return to find their coffee had changed flavor, drastically. Tabasco sauce was his favorite ingredient. He also used a very fine string attached to a spoon or a dollar bill which he placed on the floor of the café. When someone stooped over to pick it up, he would jerk the string.

Marshall had a friend who worked at one of the Columbia River dams. They had become acquainted at Martha's Inn. His friend took his lunch to work each day. One day Marshall took a sandwich out of his friend's lunch, added some dry cow manure, than put it back. He got in trouble for that. When the Fields had company, the guests were likely to find an old cow head or a pile of dry cow manure somewhere in the car when they got home. He bought imitation hamburgers made of plastic. Marshall had Janie put one in a bun for Eddie when they were in for lunch one day. Eddie spread mustard and the trimmings and took a bite. That burger wasn't up to par for his mother's normal culinary ability. The boys had their moments of retaliation and claimed Marshall didn't take a joke as good as he gave it.

He was a Chamber of Commerce unto himself. Often he would spot a stranger in Martha's Inn and get acquainted on the spur of the moment, telling them about George and farming. If someone was around for any length of time, they knew that tall, friendly guy with the western hat and cowboy boots.

They didn't have much time for golf or other recreational past times. However they loved attending round and square dancing sessions. They went with Wilford and May Erickson from George along with several more couples from Quincy and had a great time. Marshall and Janie were avid supporters of the Quincy Jackrabbits during basketball seasons when the boys were playing in high school. Rodeos were about the only other activity in which they indulged. They attended the National Finals twice, once in Las Vegas and the other in Oklahoma.

Marshall joined the Quincy Rotary in 1976. He loved going and hardly ever missed a meeting. He was president in 1980 and 1981. In 1996, he received the highest advancement in Rotary, and was made a Paul Harris Fellow. Eddie, his son, was instrumental in his selection of the honor. It was Marshall who suggested to the Rotary members to put signs along the highway identifying crop varieties for passing motorists. They are prevalent in the George - Quincy areas, Marshall was one of the volunteers to change them each year as crops rotate.

Marshall has been a member of Washington Cattle Producers and the Farm Bureau and found time to be on the Quincy Fire District Board for many years. They belonged to the George Community Hall. Marshall, along with help from Elmer Kniep, John and Esther Watkins, Zane Newton, Irwin Myers and others, tore down a building at Vantage. They used the material to start building the hall, which took years to complete. Janie was the secretary of the group and Marshall was on the Board. A Pay to Hunt program in the local area helped finance the building. He started the $100 Club at the George Community Hall. Its purpose is to raise funds for the large fireworks display every 4th of July at George.

Marshall smoked for many years and had emphysema. He came in the evening of October 26, 1999, commenting about having trouble with his breathing and used his breathing machine for that purpose. About then, five loaded trucks arrived. Out he went to help Kelly unload all the incoming cattle. The next morning he passed away. Many people have since told Janie, "I'll especially remember Marshall because he always made me feel welcome."

Marshall at Luke Field, Phoenix, Arizona - January 1945

Janie Ertel in last year of Nurses training at Colorado Springs, Colorado. 1944

Right - Janie in nursing school at Colorado Springs, Colorado - 1943
Right - Marshall pilot training in Oklahoma

The new Nash auto - Yellow Jacket, Colorado

Marshall's second love

Arthur, Addie, Janie, Robbie, Marshall
Kelly, Eddie and Marsha

LeRoy and Cleo Gossett

Texas To Washington

    Young men pursue prospective mates in many ways. Some send flowers, some sing lullabies, others write poetry or try to impress them with candy or even lavish gifts. One shy young man proved there's always a different approach to attract the object of his affections, by using a paper punch.
    The GI Bill was the primary source of funding for LeRoy Gossett's expenses, but he also worked part time at the school cafeteria. As cashier, he punched the meal tickets of the diners as they went through the line. In 1947, there were only 800 students enrolled. He noticed one young lady in particular as she passed through the line each day. It wasn't long before he found her name was Cleo Johnson. She finally became aware of his interest when she realized her meal ticket was lasting much longer than normal. Of course he maintained she came through the line flirting with him. At any rate, she finally accepted one of his requests for a date and they became better acquainted and a new friendship that would last a life time began.
    LeRoy is soft spoken and mild mannered. He has a ready smile and always a friendly word for everyone who crosses his path. He

looks the way a classical Texan should, with pale blue eyes set above a firm, square-set jaw. He stands ramrod straight and although he left Texas 50 years ago, Texas hasn't left him. His speech still has a trace that leaves no doubt the whereabouts of his origin.

LeRoy was born in 1927 at Herford. A town with a name like that could only be in west Texas. He was raised on a dryland farm. Horses were the power source for working the fields. They raised cotton, peanuts and corn. He was the second oldest of three brothers and one sister.

His family didn't own a car until just before he graduated from high school. After pleading his case with his parents for what seemed a considerable amount of time, they bought him a bicycle which he used for transportation. Girls weren't on his mind, the bicycle and playing basketball were his main interest in his high school years. The farm was 14 miles from the school. He rode the bicycle to a friend's home whose family had an automobile and rode to school with them for the ball games.

He graduated second in his class from Slidell High School in 1944. There were only 16 graduates, so he said there wasn't a whole lot of competition. The same year his family moved to Portales, New Mexico for a short period and than back to Wise County, Texas.

He attended Texas A & M University for one year before entering the US Army Signal Corp in 1946. LeRoy said he learned more discipline in ROTC that year in college than in the army. He was stationed near Honolulu in Hawaii for a year. It was pleasurable duty just after the end of the war. For a time they could check out a jeep on the weekends and tour the Island. In 1947, after 18 months in the service, they discharged him at the rank of buck sergeant. He then entered New Mexico AM University.

Cleo Johnson was born at home in the rural community of Cottonwood near Artesia, in southeast New Mexico, about 40 miles from Carlsbad Caverns. Her folks were involved in irrigated farming near the Pecos River. They used water from artesian wells where the water flowed freely out of the ground. Her dad raised cattle, cotton, alfalfa, milo and kids. She was the twelfth of 14 children, nine girls and five boys. Seven of the girls went to New Mexico State and six of them married Aggies. LeRoy said her father couldn't find enough boys around home so he sent them to a school where they were abundant.

All of the Johnson children worked on the farm. Hired help was scarce during the years of World War II, but with the help of 14 children, the family got the work completed. Cleo was driving a team of horses in the field cultivating by the time she was 12 years old. She also operated a buck rake to bring hay to a stationary hay baler. Cleo and her sister punched wire on the baler at an early age. She sat on one side of the baler and her sister on the other. Her sister poked the baling wire through from her side and Cleo pulled it through and punched it back farther down the baling chamber. Her sister would then tie the ends together. A tied bale was complete when it pushed out the end of the compression chamber. The men fed the hay into the baler with pitchforks, then put blocks with grooves in them to separate the hay in the compression chamber so the girls could thread the wire through.

 After the war, when new equipment became available, Cleo ran a tractor and side delivery rake. She raked the alfalfa into rows for the pull type balers that picked the windrowed hay off the ground. She also learned how to cultivate with the tractor. Along with her sisters, Cleo spent many summer days hoeing Johnson grass out of their fields. Her father, to their delight, replaced them with another crew. He discovered geese loved the Johnson grass roots. They started with a few geese and they multiplied and put the girls out of one job. Cleo can still remember her dad watching the geese nests for eggs and sprinkling them occasionally with water to enhance hatching.

When Cleo's mom returned from a trip once, her dad surprised her with a 100 baby chicks. Eventually they would become fryers. It was Cleo's job to watch a kerosine brooder heater burning in the small brooder house. The heater was quite old and was prone to malfunctioning, so she checked on it regularly. One day she opened the door to find straw around the heater on fire. Cleo tried to herd the baby chicks and a dozen goslings out the door, but they all ran to a corner in a huddle and she couldn't get them to move. By that time the fire was getting out of hand and she was forced to leave the shed and call for help. Her folks were away at the time, but the rest of the family saved the nearby buildings. The brooder house and its contents burned to the ground. Night was approaching when their parents got home. When her dad came in the door he greeted them with, "Well, how did things go today." One of the girls spoke up and said, "Well, we kept the home fires burning."

Cleo is a soft spoken lady with many talents other than handling a team of horses, threading wire for hay bales and fighting fires. Besides running the farm and raising three children, she is quite an accomplished artist. Her painting of Jesus is displayed along with a tapestry in the Ephrata First Presbyterian Church. Her preparation for life came through the family chores as a child and she said the Extension Service was a big help while she was growing up. Her mother and sisters were all involved. They learned how to can foods along with other household and garden chores that were handy to know later in life.

By the time Cleo left home, the artesian wells on the farm quit flowing freely because after the war many other people started farming in the area. The extra draw of water dropped the water table and they had to install pumps for their irrigation.

LeRoy graduated with an Agricultural Engineering degree and a new wife in 1950. Even though he had a degree in AG Engineering, LeRoy still had farming in his blood. The newly married couple moved to New Mexico where his dad was still dryland farming. The primary crops were maize and sorghum. They were now farming with tractors instead of horses. Being dryland, they made corrugates and planted the seed in the bottom to reach the moisture for sprouting. With the plant in the bottom of the corrugate, it was very susceptible to being covered with sand if the wind blew hard.

They were on the farm a few years before LeRoy said he figured out why it was called dryland farming. Severe droughts were common between 1953 and 1957. He never did like irrigated farming, but the crops and management are more dependable than dryland. He said nothing was more frustrating than a crop growing a couple of inches, get a sand storm and have the plants covered up. The only thing they could do was sit and pray for rain and enough moisture to replant it again.

By the end of 1953, Cleo didn't have to help with the dryland farming, instead, she had her hands full raising their three children. Wayne was the oldest, then Dave and last was Diane. LeRoy quit farming and started working for the Soil Conservation Service in New Mexico. He became an Area Engineer and worked for seven years until 1962. He applied with the Civil Service Registry and through that had an opportunity to work for the Bureau of Reclamation in Ephrata, Washington. They really didn't know much about the project until they arrived at his new job. They just took it on faith that everything would work out wherever they went.

At the Bureau he worked in the office and didn't have much contact with farmers or spend time outdoors. His duties were developing waterway designs, laying out laterals, structures, drainage, culverts and the overall distribution systems to get irrigation water to the land in the most efficient way possible. The last few years with the Bureau he spent in the Drainage Division. He worked on plans for the underground tile placement to drain fields of high water tables. The high water tables appeared in some areas resulting in salts coming to the surface. They installed the drain fields to reclaim the affected land.

The Gossett family lived on the corner of Roads E and 7 from 1962 through 1965. In 1964 they bought a quarter section of ground a mile south on E Road. The area they acquired was very rocky. They spent many hours removing all sizes of rock from the fields. Some by hand and others with a mechanical rock picker drawn behind the tractor and operated like a potato digger. Fifty acres had been leveled and farmed with furrow irrigation before they acquired it.

They thought the work and animals on the farm would be good for the children. Since Cleo was an experienced hay farmer from New Mexico, she would be a natural to run a small alfalfa farm while LeRoy drew wages in town. The boys were still quite young, but old enough for Wayne to drive a tractor for the neighbors during hay harvest. One day, Dave was sitting on the fender of the tractor when Wayne, starting from a stop, popped the clutch out too fast causing the tractor to jerk, throwing Dave off in front of the rear tractor tire. Before Wayne could stop, the wheel ran over Dave. When they got home, they told Leroy and Cleo they had run over Dave, he was dirty and had tire-tread marks on his back, but otherwise unhurt. He had landed in a shallow depression and didn't get the full weight of the tractor wheel.

The salary from the Bureau of Reclamation allowed them gradually to develop their farm without going into debt. Many early farmers had to acquire loans to get started, that is, if they had credit. They bought the ground from Harold Beltz. There was a house on the property at the time of purchase. It was unique since it was built from hay bales.

In the house made of bales was an organ and piano along with a pot-bellied stove. A large wool rug was on the floor and original oil paintings on the walls. Knotty pine covered part of the walls. It was a two room structure, a living room and bedroom. There was no bathroom, at that time it was an outdoor privy instead. But all in all, quite cozy. Outdoors, there was a metal tank suspended high enough to stand under. The sun heated the water during the day and with canvas walls on the sides and a ditch underneath to catch the water, it made an adequate shower stall. They may have thought the winters quite long when it couldn't be utilized for months at a time.

They bought a small house in Ephrata and moved it over the top of a full basement on their new farm. It replaced the house constructed of hay bales. They purchased another 160 acre farm in 1983. It was directly north of their home place, giving them a half section in one piece. The new ground originally belonged to an Ephrata attorney, Cliff Collins, who sold it to Vance Burke until Gossetts acquired it.

The children were active in 4H and FFA projects. Dave and Diane both were honored with a trip to Chicago for the 4H Congress which was a national event. The honor wasn't new to the family as Cleo attended the national meetings during high school many years ago.

A pasture of orchard grass was planted for livestock. Small Holstein and 500 pound feeder calves were bought in the fall, wintered over and put out on pasture during the summer. The weekends were spent working cattle, giving needed shots and other necessary attention. "Buying cattle at 34 cents in the fall, feeding them hay all winter and pasturing them all summer, then selling for 28 cents a pound wasn't very profitable", Leroy said.

Cleo spent many days chasing cattle while Leroy was working in town. She was raised around cattle, but she said, "Holsteins are the dumbest creatures there ever were, I couldn't get them through a gate by myself, I always needed help." Their neighbors, the Moritzs, helped a frustrated Cleo put them back in several times. The hay stacks had to be shovelled off when it snowed. LeRoy fed them in the dark before going to work and again in the dark after returning. After a batch of cattle sold, Leroy figured out how they came out financially. It seems they were $90 to the good. Cleo finally rebelled. "I could take in sewing and make better than that."she said. They both decided they wanted to farm pretty bad to go through all that trouble with so little return. They quit raising cattle and concentrated on hay and wheat to sell on the market.

They started with furrow irrigation on the 50 acres, but as time passed they expanded the acreage and went to handline sprinklers, then to wheel lines and finally in 1980 they put on a circle pivot to cover 130 acres of their quarter section. Before LeRoy left in the morning for his job, he decided what needed to be accomplished that day and Cleo had the boys carry out the tasks at hand. Each hay cutting, when the baling was completed, Cleo drove tractor, pulling a slip while the boys pulled the bales onto the flat surface. When full, they went to the stack yard where the hay trucks picked up the alfalfa for a destination west of the mountains. Dave didn't like the position the slip was in at the stack with his mom driving, so after the slip was loaded and ready to go to the stack, he drove and Cleo rode on the slip. They didn't think much about it until a few days later a friend told her all the neighbors thought it was awful that Cleo had to load that hayslip while that big kid drove the tractor.

Over the years they stuck basically with alfalfa and wheat because their ground was rocky, making it difficult to raise row crops which require cultivation. All three children grew up and attended Washington State University. Dave graduated with a Masters in agricultural economics. He worked a short time as an extension agent in Ephrata before deciding in 1977, he would rather farm than advise others how.

LeRoy had a background in the Baptist church and Cleo came from a Methodist family. The year of 1972, they started going to the First Presbyterian Church in Ephrata. They have been very active over the years with Cleo serving time as a Deacon and LeRoy as an Elder. After renting the farms to Dave in the late 1990s, they have had time to do volunteer work at Cannon Beach Conference Center in Oregon for short periods. Both of them have enjoyed meeting new people at the non-denominational facility located on the shore of the Pacific Ocean.

For their 40th wedding anniversary, they traveled to Hawaii. While there, LeRoy tried to locate his old barracks southeast of Honolulu at Fort Shafter. The only problem was things had changed in 42 years. A freeway now went through part of the area. The main gate was still intact, but many of the original buildings were gone. He felt rather dated when he asked a fellow nearby about the location of the old base buildings, the reply indicated he knew nothing about them; he wasn't even born yet when they were there.

On their 50th anniversary, her twin sisters asked Cleo if she remembered when they were girls on the New Mexico farm how they would lay down in the irrigation ditch when there was a ditch break and have soil shovelled up against them to stop the out flow of water until the hole was plugged? Things sometimes don't really change that much over the years.

Other than trips back and forth to visit family in New Mexico and to visit their son Wayne northwest of Phoenix, AZ. They haven't travelled a great deal, except once in California they took a commuter plane to a larger airport to return home. It was a twin engine propeller driven aircraft. They were the only passengers on the flight.

They were nearing San Francisco when the co-pilot jerked the curtain closed between the passenger cabin and the cockpit. They could see a red light flashing on the panel as the curtains were closed. About that time, one of the props quit turning. Needless to say by that time they were wondering what was happening. After they landed at San Francisco they noticed fire trucks and ambulances were standing by the runway when they landed. The plane stopped way out on the ramp away from the loading area. As Cleo was getting off the aircraft, she started to retrieve her overnight bag, a crew member hurriedly said, "Don't bother with that, hurry, just get off, just get off."

As a ground crew member took them to the terminal in an auto he told them, "Oh, that happens all the time with those old airplanes." When the light appeared on the panel, the pilot had shut the engine down purposely and the emergency vehicles on the runway was a precautionary measure. It turned out to be a malfunctioning light. Perhaps that's why they don't travel all that much.

Wayne and his wife, Renee, own and love horses. Wayne makes a living shoeing horses in the area and his wife rides professionally as a barrel racer. Dave and his wife Sharon, farm the family ground and rent other ground to raise potatoes and alfalfa south of Ephrata. Their daughter, Diane, lives in Seattle where she worked in real estate and then became a ballroom dancing instructor. They have seven grandchildren, Dave has three, Diane three and Wayne a step daughter.

LeRoy had a good eye to pick out the girl in the cafeteria line that really fit the bill for what the future had in store for them. Never underestimate the effect an inanimate object may have on romance, even a handheld paper punch.

The Rock Picker

I.P. Johnson Family Christmas 1951

George, Charles, Lamar, Herman, Bill
Helen, Libby
Edna, Chris, Cleo, Lorene, Mary
Delta, Mom, Dad, Neta

LeRoy's family in Texas when he was 16 years old

LeRoy in the Army

Cleo at 13 years old

LeRoy and Cleo Gossett on their 50th Wedding Anniversary
1997

Cleo and her eight sisters

The house of straw bales

Wayne, Cleo, Diane, LeRoy, Dave

1967
Wayne, Diane, Dave
LeRoy, Cleo

Bob and Virginia Graham

A Cowboy In Work Shoes

It was October 1953. The day was warm with a bright blue cloudless sky covering all of Central Washington east of the Cascade Mountains. It was a beautiful day driving from Rock Island east through the Columbia Basin toward Moses Lake. It was a day of surprise for Bob

Bob Graham had been this way before. He was used to the long grade called the Trinidad hill coming up from the bottom of the Columbia River canyon. He had an overhauled engine in a dark colored 1946 Chevrolet Stylemaster. Overheating was no problem at all and he enjoyed the many curves on that winding and narrow highway. Once over the summit, came the flat sage brush-covered countryside with the road leading straight east to Quincy and beyond.

Bob was wiry and slim. He was 28 years old with brown hair and blue eyes with more than a little devilishness in them, if you looked closely. A pert young lady by the name of Virginia, who also happened to be his wife, was beside him. Virginia had blue eyes and auburn colored hair and a forgiving nature. The forgiving nature is mentioned only because Bob was a prankster and probably needed forgiveness quite often

Bob was born in Oklahoma, he moved to Washington at the age of 19. At 15 years old, he fell nearly 25 feet out of a tree one night. He landed upright, hurting one knee that has bothered him ever since. He liked the outdoors and horses.

At 18, he was bucked off a horse and again ended up with a life long physical problem. However, it didn't stop him from working part time as a cowboy, a cat skinner and horseshoer. By 1953, he was living in Rock Island, Washington and working for a power company.

Bob had his father-in-law and a neighbor with them when they topped the Trinidad hill, they saw where acres of sage brush were in process of being torn out. Wide strips of ground were being leveled. All he ever saw there before was sagebrush and jackrabbits.

They soon learned the Bureau of Reclamation was excavating the strips for water canals. The canals would carry water to a thirsty land from Coulee City south to Ephrata and Quincy. The earth was ready to bring forth great varieties and quantities of grain, hay, vegetables and seed crops; all it needed was precious water. He hadn't even heard of the project before that time and was amazed by the construction as they drove on through the little town of Quincy.

Little did Bob realize that six years later he and Virginia would move to Quincy and work for the Grant County PUD.

He eventually worked at Priest Rapids and Wanapum Dams on the Columbia River. They had lived in Concrete, Washington, where it was always wet or foggy. Bob and Virginia disliked the almost constant high moisture and rain. One night Bob discovered himself driving on the sidewalk because the fog was so thick he couldn't see the street. That prompted a move to the east side of the mountains and a drier climate.

Bob and Virginia found a house in Quincy when they first arrived. Houses were scarce with all the construction going on. Families were continuing to arrive and finding a place to live was a luxury. To be closer to the job at Priest Rapids, they again moved to Mattawa where the PUD had moved some houses from the Tri-City area for their employees. The houses were not on foundations. The buildings were just set down on the ground.

At Mattawa their two daughters went to a small country school. Virginia said the girls had never been to a school where the girls wore jeans. She was a little upset and stated her girls were NOT wearing jeans to school, they would be in dresses. It wasn't long after that proclamation when the wind storms started. A pile of sand several feet high was between the house and the street. After one bad wind, Bob swears half the pile moved into the house. They had a blue daveno and couldn't tell what the color was afterward. Every thing in the house was buried in sand. Bob said, "Even I felt like bawling." That's when Virginia decided the girls could wear jeans to school after all.

During the 1960's the school district wanted their children to attend the Red Rock High School in Royal City. Evidently the new school didn't have state accreditation. The Grahams decided they preferred Quincy which was further away, but worth the effort to get them there. They found a high school boy that also wanted to attend Quincy. With their permission, he drove Graham's car back and forth with a full carload each day.

The winds were especially bad in the area along the river. Bob said one eastbound train had 17 flat cars blown off the bridge while crossing at Beverly. It dumped loads of stereos and shoes into the river. Some of those cars are still in the river just below the Beverly bridge. The construction site at the dam also had its share of windy incidents. (I don't mean Bob.) He was working at Wanapum Dam when the wind blew metal buckets and sheets of galvanized metal through the air. As they passed the boxcars in the switch yard, it created lightning in the immediate area before they blew on over into the old bombing range toward Royal City. That area had all kinds of loose debris including partial roofs of houses.

During the first years, Bob said the new farmers lived in trailer homes and shacks. Some were in sheds we would now consider a chicken coop. A few came with enough money to build decent homes, but not many. The main north-south highways and east-west highways were paved. Later they paved the Dodson and Adams roads. All other roads were gravel, dirt, or nonexistent.

The dust storms were many and lack of visibility caused many accidents. At the corner of Road 5 and the Quincy-George highway each spring, the owner worked the ground up so fine it would blow with any wind. One wreck involved nine vehicles and one person was killed. Another time at the same spot during a blow, it was zero visibility. Bob had to pull over and stop when he felt the front tire going into the barrowpit. One welding truck just barely missed him. The truck must have been going 60 miles an hour and couldn't see a thing. A few minutes later Bob did get hit. A car side swiped him and kept going. He set there a minute before deciding if he stayed there he was going to get hit again. He eased out and within a few hundred yards was clear of the blowing sand. There sat the car that hit him; it happened to be the farmer's wife who owned the ground.

Bob worked ten days and off four. On the off days he worked for Stuart Scott and other farmers that needed help with farm work. He did horseshoe replacement, and electrical repair. Bob drove potato truck for Holloways during potato harvest every year after retiring from the PUD. He enjoyed himself immensely the times he could pull some thing on the storage crews. One day he let a dead and very ripe skunk unload out of the truck onto the conveyer where the crew was sorting. He laughed while they scattered in all directions. He threw Earl Holloway's hat in the air once and tried shooting holes in it with his pistol.

One harvest in particular, his knee was hurting more than usual. When the crew noticed him limping, Cathy Flint, who was helping that year, asked him why. Bob told them his pegleg was giving him trouble. The crew felt sorry for him until his daughter came by the following week. Someone asked her about his leg and how he drove truck with it. "He doesn't have a pegleg" she said, "He's just pulling yours."

Bob and Virginia raised two boys and three girls in the basin, retired from the PUD, and now live in Ephrata. When asked if he could do it all over again, would he still come to the Basin? Bob answered. "To get away from the coast and the rain, yes." When they asked Virginia the same question, her answer was, "Yes, I like it here, but I wouldn't go back to Mattawa."

Bob riding in the Moses Lake Rodeo in 1944.

Dean and Russella Hagerty

I'll Go Where You Go

The Allison Iowa High School didn't have a football team, but they did have a band and an extra E-flat tuba. Dean Hagerty, a slightly built and energetic teenager, who loved to play basketball and baseball, lived across the street from the band director. Two weeks after being recruited for the band Dean found himself filling in for a missing tuba player at the spring music festival without a mouth piece for his instrument. The director took it away to make certain he didn't screw up the performance.

The missing tuba player was one of twin boys who played tuba in the band. One brother became gravely ill and died shortly after Dean joined the band. It wasn't long before the director consented to let him perform with his instrument intact. Before high school was over, he did learn, very well. He won two or three contests after getting the mouth piece back and a lot of practice.

Dean was born in Sibley, Iowa, in the year of 1925. When he was five years old, his parents, two brothers and one sister moved to Allison, Iowa. It was a small town in a farming community. Work was hard to find during the Great Depression. The family was fortunate because his father ran a lumber yard. Young people were allowed to work during those years. Dean remembers unloading coal out of a gondola railcar when he was 16 years old. The coal, he remembers, came from Roslyn, Washington.

Dean was in his senior year when Japan attacked Pear Harbor December 7, 1941. After graduation the spring of 1942, he went to work at Rath Packing Company, a hog processing plant in Waterloo, Iowa. The fall of that year, his friends and neighbors selected him, via the draft board, to join the armed forces of the United States. Company 830 at Camp Benion in the Farragut, Idaho Training Center was his new home when starting his hitch in the U.S. Navy. His time spent learning to play the tuba finally paid off, instead of KP and marching, he was playing in the band. To the envy of his buddies, he was also assigned the duty of a lifeguard at the swimming pool Thursday nights. Thursday was the nurses night in the pool.

From Farragut they sent him to the University of Illinois to learn diesel engines. Upon graduating, they shipped him to the navy pier in Chicago for advanced diesel school. Returning to Farragut, he was assigned to a mine sweeper, AM 306, an all-metal ship.

Winifred Russella Thompson started life's journey at her parents home near University High School east of Spokane October 24, 1926. She was born in the family's upstairs bedroom, so she came into this world feeling right at home. Russella attended Opportunity Grade School and then West Valley High School at Millwood, Washington. Her father was a carpenter-contractor and built homes in the area. Her mother raised the girls and cooked at a nursing home housed in the women's dormitory of the Christian University. The university later went bankrupt during the depression years.

When Russella was 16 years old, she and two of her sisters worked in the same nursing home as their mother. She had a natural instinct for serving and caring for people and enjoyed interacting with the elderly patients and thought she did well with them. The experience helped mold her desire to become a nurse. However, there were some terrible conditions and treatment, including physical abuse by the owners of the facility which distressed her.

Roller skating was popular with young people during the 1940s. On January 10, 1944, Russella, then a senior in high school, went with friends to the Monterey Skating Rink in Spokane. It was located across the street from the Deaconess Hospital. The rink was later torn down. They built interstate 90 where the building once rang with music from the loud speakers heard above the rumble of hundreds of spinning skate wheels as they flew across the floor.

Russella was at the skating rink that evening. She saw a slim, handsome young sailor come through the entrance. She was instantly aware of his presence. She was upset because the "lady's choice" number was over and she didn't know how to meet him. Evidently, the attraction was not one sided because it wasn't long before the handsome young sailor nervously asked her to skate with him. The sailor and his buddy were on a weekend pass from Farragut, Idaho Naval Training Station. After the rink closed for the evening, the young sailor accompanied Russella, her two sisters and a girl friend to the Greyhound bus terminal for their return home out in the valley. Before the bus left, the sailor arranged to meet her the following day in Spokane to spend the day together.

Every time Dean was able to get off base, he took the Spokane bus and walked a half mile to Russella's home. They corresponded between visits. Russella thought he was a tall and handsome fellow, and his shaving cream smelled so very good. Dean never did ask her to marry him, it was somehow understood that it was meant to happen. They became husband and wife on August 28, 1944. The marriage ceremony took place in the Thompson home at 10:10 a.m.. Dean caught a bus by 2:00 p.m. for Seattle where he was to report at his new home away from home, the USS Spector.

A week after Dean reported to his ship, Russella went to Seattle and found a one room apartment in a private home. Dean came home every night for about two weeks before the Specter and its crew sailed for the South Pacific and the war with Japan. Russella wrote a letter every day. She marked each letter on the back with SWAK (sealed with a kiss) and punctuated with red lipstick from a kiss.

When the Specter headed overseas, Russella returned to Spokane. It had always been her dream to be a registered nurse. Nursing was the only occupation she ever wanted. She entered Sacred Hearts School of Nursing as a member of the Cadet Nurse Corp. The cadet nurses were a way of releasing registered nurses to go into the service. The government paid the cadets a small salary as they trained. It took three years of training to become a registered nurse. Meanwhile, Dean was sailing out of Seattle aboard the newly commissioned ship commanded by Lt. Jacques Chevalier, USNR. They set sail September 16 for training off San Pedro and San Diego for a month before departing for Hawaii in November. Their training consisted of antisubmarine warfare, gunnery and minesweeping operations. The Specter and crew left Hawaii the 22nd of January 1945, for the Volcano Islands.

After stops at Eniwetok and Tinian, Specter arrived at Iwo Jima on 16 February, three days before the landing, and began minesweeping operations. They were busy clearing minefields, patroling and performing escort duty until 28 February, when they sailed for Saipan. After a stop at Ulithi from 6 to 19 March, Specter steamed to Okinawa. On arrival there on 25 March they commenced preinvasion minesweeping operations of sea lanes and channels. The ship remained in the Okinawa area until 6 August. During this period, they conducted antisubmarine patrols, swept mines off of Okinawa and Iheya Shima and made two open sea sweeps in the East China Sea. The Specter's crew shot down a Japanese fighter plane while on patrol off Le Shima on 25 May.

They were off the coast of Iwo Jima sweeping for mines three days before the U.S. troops landed. They were close enough to shore that Japanese riflemen were shooting at them without any effect. The enemy wouldn't fire the larger artillery at them because they were saving all the firepower possible for the coming invasion. During the invasion, the Specter was just offshore when the famed flag raising took place on Iwo Jima. They were near Kermaretta Island south of Okinawa the night the famous U.S. war correspondent, Ernie Pyle, was killed. A kamikaze pilot hit a nearby ship, the USS Swallow, at 6:00 p.m. By 6:15 it was bottom side up and at 6:30 it sank below the surface. The Specter picked up 30 or 40 survivors. Dean remembers the flesh hanging from some sailors brought on board, he doubts that many of those badly burned men survived for long.

A Kamikaze also hit and badly damaged a hospital ship, the USS Hope. When they left Iwo Jima, the Specter, along with two other minesweepers escorted the USS Hope back to port. The minesweepers were very powerful, but slow. The USS Hope had to zig zag all the way so the minesweepers could keep pace. They called the Kamikaze organization "The Divine Wind." It was a great honor for a Japanese to be a Kamikaze pilot. During the battle at Okinawa the Kamikaze hit over 300 U.S. ships, killing thousands of personnel.

The Specter had a Dalmatian Coach dog as a mascot. On a morning in 1945, they were in the East China Sea. They received a radio message from another ship saying it picked a dog out of the ocean, did anybody have one missing. The crew of the Specter checked and found their dog, Spec, indeed was nowhere to be found. The dog was returned unharmed. There was some speculation on how Spec went overboard. They made no accusations, but everyone knew that Spec always did his thing in the gunnel where the gunner's mate had to clean up after him. So they knew who tried to get rid of the dog.

Thirty six years later while on a Chamber of Commerce tour of the Port of Seattle, Dean was seated next to Seldon Jones from Moses Lake. As they were cruising along, Seldon said, "Hey, Dean, there's a ship like the one I served on in World War Two, that's a minesweeper." They started talking about their service time and Dean told him the story of Spec being lost over board. Dean noticed Seldon looking strange and he finally said, "Dean, you are not going to believe this, but I'm the one who jumped into the ocean and picked that dog out of the water."

The Specter was in Leyte Gulf in August. During this period, on August 6, the first atomic bomb was dropped on Hiroshima. Three days later, a second bomb was exploded over Nagasaki. After almost four years, World War II was over when Japan surrendered August 14, 1945. The Specter sailed for Japan on 28 August and, after checking in at Buckner Bay, Okinawa, arrived in Japanese home waters on September 9th. During the next three months, they swept mines at Nagasaki, Sasebo, Bungo Suido and Tsushima. On December 11, 1945, Specter was ordered home; arriving at San Diego on January

11, 1946. Specter received four battle stars for World War II service. The ship and crew set the record for destroying more mines than any other ship in the navy, a total of 462 during a 90 day period.

Giving up nurses training was very difficult for Russella before the three years were completed to become a registered nurse. She wanted to meet Dean in Iowa when he returned from the South Pacific. She cried a good deal before catching a train to Minneapolis where he was to receive his discharge. When getting off the train she thought she was going to die because it was so cold. It was twenty degrees below zero. Russella wasn't prepared mentally or physically for the temperature. A light cotton coat was all she had to ward off the horrible penetrating chill. It had to be true love.

On January 20, 1946, Dean received his discharge from the navy at World Chamberlain Airfield, which is now the Minneapolis-St. Paul Airport. It had always been Dean's ambition to go into business for himself. Russella and Dean moved to Allison, Iowa. The population of the teaming metropolis was almost 500 people. Russella had saved her wages while in the Cadet Nursing program and had saved all the money Dean had regularly sent back while in the navy. Dean had his mustering out pay and they received a G.I. Loan to set up a machine shop of their own. There was another established blacksmith in town. They were on the short end of the stick, being the new kids on the block. After several months, Dean concluded the area was not populated enough to support two shops.

They disposed of the shop and Dean took a job running the City Service Station. It wasn't long before plans were made to tear down the service station to make a parking lot. In 1947, Russella's sister and husband, who lived in Dayton, Washington, offered Dean a job managing a Phillips 66 Service station and four tourist cabins. Dean accepted, and they caught a bus to Minneapolis where they could fly west to Spokane. The timing was not the best as Russella was expecting a baby in two weeks. In Minneapolis, they watched a centennial parade during the day and stayed the night. Russella had a terrible backache during the night and thought the trip had been too rough on her. Before the night of July 25th, 1947 was over, she was in the

Minneapolis hospital with a new baby boy, Russell Dean. There was a two-week delay catching the plane to Spokane and Dayton.

Russella's maternal grandfather's name was Jacob S. Rainwater. He came from Tennessee, following a brother to Oregon. He made two or three trips across the country before finally settling in Dayton, Washington. His first wife died and he remarried a nineteen-year old young woman and raised a second family. He died in an influenza outbreak leaving the second wife with 10 children ranging in age from two to twenty-one years old. Russella's mother was born in 1889 at Dayton on the homestead. Being the fourth oldest, she had to quit the 8th grade to help raise the family while two older brothers worked the farm. It was the same year Washington gained statehood. Russella's cousins still live and operate the farm on Robinette Mountain.

The family gave an old house on the homestead to them. Dean cut it in half, moved it onto a lot in town and reassembled it for a place to live. The house was so old it didn't have studs and was insulated with newspapers. A daughter, Claudia Lynn was born in Dayton July 30, 1948.

It didn't take long for them to decide there wasn't much future for them in Dayton. Russella suggested Dean enter college on the G.I. Bill which would expire if not used. On short notice, the only college he could get into was Eastern Washington at Cheney in the fall of 1948. The winter of 48-49 was a bad one. It made travel difficult and sometimes impossible between school and Dayton where Russella stayed. That was not a satisfactory arrangement, so Dean transferred to North Idaho College of Education in Lewiston. Housing was scarce, but the football coach at North Idaho came from Dayton and knew the Hagertys. He informed Dean that if he turned out for football, he could get him a place to stay. Dean had never played football before, but he learned, and they had a place to live. He also got a part time job at Chapin Service Station and Auto Parts. They had a contract running a school bus which Dean drove in the evenings.

Governor Jordan of Idaho closed the college during their second year of school and they moved to Central Washington College at Ellensburg where he finished his schooling in 1953. Dean worked at a Napa Auto Parts before and after graduation. He become acquainted with a farmer, Fred Davidson, who was a customer. One day Fred came in the store and asked him what he planned to do after graduation. Dean informed him he had signed a contract to teach school at Grandview, Washington. Fred asked if he ever thought about going into business for himself? Dean said, "No, he really hadn't." Fred replied, "Well, if you ever do and need some money, come and see me."

They moved to Grandview where Dean taught school for a year. Dean and a friend from college, Glen Rayburn, who also was teaching at Grandview, thought the teaching salary of $3600 a year was really not meeting their needs. Dean and Russella had three children by this time and another, Robert Wayne, born August 30, 1952, on the way. Dean thought they should be doing better financially. Russella thought they were doing O.K. and liked having Dean in the teaching profession. The two of them had quite a discussion about their plans and she thought it was settled he would not go into business, but would return to school for a Masters degree. That was before Glen, Orville Clough and Dean had a meeting and decided they would like to start their own parts store.

The three entrepreneurs drove to Connel, Othello, Warden and on to Moses Lake looking for prospective sites to fit their plans. When they stopped at an intersection in Moses Lake, Dean saw Gib Kaynor, who had been the bookkeeper at the parts house in Ellensburg. Upon seeing and recognizing Dean in the car, he called out "Come on in and lets have some coffee." They all went into Elmer's Café. When they sat down in a booth, Gib introduced them to another patron, Mode Snead. They said they were looking for a place to start an auto parts store. After visiting awhile, Mode said, "If you're really serious about this, I'll move my John Deere business into half of my building and you can move into the other half."

Dean contacted Fred Davidson, the farmer friend near Ellensburg. Fred said, "I can trust any man who has a wife and two children and still go to college. He has to have a lot of savvy and guts to accomplish that." He loaned Dean $5000 to get started. Years later when Dean became a Grant County PUD Commissioner, he found that Fred had been the fish biologist for Grant PUD at the time he loaned the money.

The doors of Columbia Automotive opened in 1954. The same year another daughter, Marie Clarice was born on August 10th. Irrigated farming was just getting started and there was a great need for the service they supplied. As the farming community grew, so did the auto supply business. Eventually, Mode Snead built a new building for his own business and the Auto parts took over the extra space in the original location. The store was as much a part of the agricultural scene as the tractors and trucks it helped to keep running. They opened another store in Ephrata, then in Brewster. While the second Bacon Syphon was being constructed, they opened one in Coulee City. The Ephrata store wasn't doing too well and they moved it to Pasco where it remained for 20 years. Since they were close, they opened one in Kennewick and another in Richland. The stores expanded to Hermiston, and Milton-Freewater, Oregon and Walla Walla, Washington. During that period, on August 24, 1958; Yvonne Louise, their last child was born.

The expansion took its toll. Dean spent all of his time on the road rushing from one point to another. Finding qualified people to manage the different locations was extremely difficult. Too many new employees first question was, when is payday, the second was when can I have a weekend off. He missed part of his kids growing up while he drove more than a million miles during a 25 year period. As soon as he got home from one store, he would receive a call from Walla Walla or Hermiston. Someone had quit or items were being stolen, requiring him to head for another location. The burden became too troublesome and they started closing the outlaying stores one at a time. They ended up with the one in Moses Lake which closed July of 2000 after almost a half century of supplying the needs of the community.

Russella always wanted to be a registered nurse. In 1965, she finally came close to realizing her dream by attending Big Bend Community College at Moses Lake and received an associate degree as a licenced practical nurse. Later in 1980, she took a cosmetology course and became a licenced cosmetologist. Every Saturday night for many years, Russella has danced to the music of the Sundowners band at the Ephrata Senior Center. Dean and his tuba help the band keep the place jumping. He sang with the Basinaires, a barbershop group, and acted as the master of ceremonies for many of their concerts. He is multitalented. He does scrap art, weaving together odds and ends to form replicas of people. Dean has appeared in 18 of the spring musical productions put on at the college in Moses Lake.

The Hagertys have five children, two boys and three daughters. Russell Dean is in the Moses Lake business. Claudia Lynn is a teacher in Las Vegas, now retired. Robert Wayne works in Bellevue setting up operations for client companies. Marie Clarice works for U.S. West in Seattle and Yvonne Louise works as manager for Gottschalks in Moses Lake.

Through the years, Dean's natural leadership qualities has served the community in good stead. He was on the Moses Lake Chamber of Commerce and president of the Lakeview Terrace PTA. He was 13 years on the Port District and two years as President of the Washington Public Ports. While with the port district, he had the opportunity to go to Japan twice, to Singapore, the Philippines, Hong Kong and Europe on trade missions. Russella accompanied him, paying her way with their own private funds. In the year of 2000, Dean was on his 18th year of serving as Commissioner of the Grant County Public Utility District. The PUD is in the process of re-licensing the Grant County Dams with the Federal Government. His desire is to see the re-licensing completed by 2002, if not, he would like to run for the position one more time to complete the task at hand. He has been the lead commissioner in negotiations and travelled the Northwest trying to complete an agreement with all affected parties.

Dean said he attended a meeting in Denver in 1980 for water and power. One speaker was the head of the Bureau of Reclamation who said water and water rights would be big issues from that time forward in the Northwest. That statement has come true, considering the ongoing battle with former Secretary of Interior Babbitt and radical environmental groups trying to remove dams around the U.S. The next targets are the four dams on the Snake River in Washington State. If those groups have their way, it would destroy the cleanest, most renewable source of electrical power available. In addition, it would eliminate thousands of acres of irrigated crop land. It would cause major disruptions of jobs and food production. Wetland areas created by the back water, fish and many species of wildlife would be permanently wiped out.

The environmentalists and government agencies are using the Endangered Species Act to force millions of dollars on salmon recovery studies and programs, most of which have been a waste of money. Power customers are paying up to 25% of their monthly bills on salmon recovery efforts. The ultimate goal of the environmentalists is to control the water and turn as much of the area back to the natural state before people appeared on the scene. The result would devastate the Northwest. The clean power, food production and job creation would disappear. It would do away with all that Dean and so many others have labored so hard to achieve, turning the desert into a garden.

Dean said with conviction, "Water is the only natural resource we have in the Columbia Basin and the Columbia River provides it through the irrigation districts. Our whole future depends on water, without it, we're back to jackrabbits carrying waterbags. We can't afford not to put whatever resources we need together and make sure water and the dams stay in place."

Dean and Russella's story is really a love story. When asked if he would decide to come to the Basin again, Dean said, "I had a burning desire to go into business. Yes, I would do the same thing again, the Basin has been very good to me and my family." Russella said, "I would miss the mountains and trees, but I would follow you."

Russella, 18 years

Dean, 19 years

The U.S.S. Specter AM 306
Commissioned August 30, 1944.
Lt. J. Chevalier, USNR in command

Dean in 1999

Dean and Russella
1998

Tub and Wanda Hansen

Really and Truly

Frank had led a hard life, things didn't come easy, he worked for everything he got. He had brown hair and penetrating blue eyes. When he wanted to emphasize a point, he would begin with, "really and truly." When he said that, he came across as being very serious and wanting to make a point. He was definitely the "tough, but oh so gentle" gentleman, and for an old time cowman, his speech was colorful, but definitely not profane.

Frank was a scrappy little fellow the hired men liked to tease, they would fill his pockets with mice to take home to his mother. The rugged workers at his dad's sawmill in Okanogan County of Washington State started calling the little two and a half year old "Tubby", because he was almost as wide as he was tall. They didn't realize they named the youngster for life and Frank Hansen was to carry the name of "Tub" to the halls of the Washington State Capitol building in Olympia many years later.

Frank was born December 27, 1913, at Olema, Washington to Thomas Alvin Hansen and Ida Agatha (Platte) Hansen. Tub's family included five boys and two girls. They moved to Withrow, Washington about 1915 after their sawmill was destroyed in a fire. Thomas operated a steam powered, stationary thrashing machine with a long belt drive. He took his machine and crew from farm to farm doing custom work, hauling grain to it with wagons pulled by horses.

The summer of 1917, Tub's father worked for a contractor who was to build a grain elevator at Wheeler, Washington. It was then the family moved to the small town of Neppel with a population of approximately 100. Tub, his two sisters and two brothers, attended a four room school house in Neppel which housed all 12 grades.

In 1925 they bought a small farm on Mae Valley Road west of the lake where his mother kept the boys busy while their father worked as a "high rigger" on construction of Rock Island Dam. Tub was 12 when he rode horseback in a driving rain to bring the cows to their new home in Mae Valley, a distance of eight miles. It was after dark by the time they arrived and the cows secured.

At the age of 16, he decided he wanted to be a cattle rancher. He worked summers hauling sacked grain and working in orchards for $36 dollars a month, $25 dollars would buy a cow. The farming meanwhile was all done with horses or by hand. In 1930 or 1931, more land was acquired and the Hansens become the first commercial potato growers in the area, long before Grand Coulee Dam came into being.

The potatoes were picked up and bagged by hand after being dug by a horse drawn implement. It was 1940 when they bought a new Farmall H tractor. The increase in horsepower prompted Tub's father to invent the first machine in the area that dug and sacked potatoes in one operation. People came from all around to watch it work.

Much of the area was farmed with water from the lake or wells until the power became too expensive. The power was available through the Washington Water Power from Spokane. Its cost was a contributing factor to the orchards in the area being left to die. However, the Hansens used a diesel tractor with a PTO belt running to a pump in the lake until the REA (Rural Electrification Act) came in, bringing affordable power. On the irrigated acreage in early times they raised mostly potatoes, which were hand picked into sacks, then loaded onto trucks and hauled to the warehouse.

Tub and his older brother, Russ, worked with their teams of horses on the two lane road that is now I-90 during the winters of 1933 and 1934. A makeshift cabin on the back of a potato truck sufficed for shelter while on the job. They went home on weekends to thaw out and get some good meals in order to endure another week on the job.

Tub said, "It was so cold in the mornings that, when we washed the table after the dishes were done, we left a sheet of ice behind. For lunch we simply rolled up the left over breakfast pancakes with whatever we happened to have to fill them with and stuffed them in our pocket."

When he was 22, Tub and a friend, Shirley Hussey, decided to try their hand at professional rodeo. It was the fall of 1935 when they drove to California to start the winter circuit in the southern states and on through the next summer of 1936. They went state to state in Tub's old Plymouth car. They found it tough to ride in the money, but usually got enough between them to buy gas and the entry fees at the next rodeo. Tub rode bareback horses and the bulls, his friend rode saddle bronc.

The out-of-state boys didn't seem to get quite as high a score as the locals and in order to buy gas they missed a lot of meals. Tub said later, he had never been so hungry in his life. Hunger pangs were severe when they were going through one western town early one morning. They found a truck delivering bottled milk and followed for a while picking up his deliveries and with some day old bread from a bakery, their stomachs were full for a change.

It was many years later when Tub was serving in the Washington State Legislature that was in session, when the speaker asked "Has anyone here ever been truly, truly hungry?" Tub raised his hand surprising a lot of people, he still remembered the year on the rodeo circuit.

The adverse conditions that forced growers to let their orchards die out in earlier years, were now wanting to clear the ground for other crops. Since the diesel tractor was not being used in the winter time, Tub used it in the winters of 1936 and 1937 pulling out trees from Neppel to Quincy. At least it wasn't all dull, he managed to fall into some thinly covered sewers and halfway into an old still.

Wanda Hill was one of 10 children. Wanda had brown hair and pretty green eyes that seemed to change with her wardrobe. She and her nine brothers and sisters were all born in their family home in northern Idaho. Two of them were premature, one girl was thought dead three times. One of those times she was prepared for burial when she yawned, stretched and was ready for life to go on. She only weighed a pound and a half when she was born and so delicate she had to be handled on a feather pillow.

Her parents had cattle and grew grain and hay of which the cattle utilized the biggest share. It was 1933 when her parents decided to move to central Washington. That happened to be the same year Grand Coulee Dam was dedicated. Tub was commissioned by Wanda's mom to haul the cattle to their new home in Neppel. Wheeler at the time, was a larger town than Neppel, which was near the lake. He was paid with a few head of cattle for the move.

Four years later Tub and Wanda were married in Wallace, Idaho. It was November of 1937 when they eloped. They were heading for Montana when a snow storm caught them, they made it to the county courthouse in Wallace just before closing time at noon on Saturday. Then they travelled back to Neppel and their small herd of cattle and a life time of work.

Tub had accumulated about 20 head of cattle by that time. While building their herd, the cash to live on was brought in by Tub and his brother custom hauling sacked wheat from the farm to the warehouse on the railroad siding. They would work a month during harvest every year. That was the year's cash income and enough to keep them going. Meanwhile whenever there were any extra dollars, it was spent on a few more acres of range land here and there.

Times were tough and the farm didn't bring much in. Tub and Wanda had an arrangement with Everett Wetzel, the owner of the biggest grocery store in Neppel. They would charge what was needed and whenever Wetzel needed meat, he called Tub. Tub would pick out a grass fattened beef, butcher and quarter it, and hang it in the store's cold storage to pay their bill and then start charging again.

In 1939, Tub and his brother Russ took out a loan to farm and they were budgeted $20 a month for groceries. They had to list and keep track of everything spent. It was a good thing they were raising potatoes and had a big garden. That was also the year the town of Neppel was changed to Moses Lake.

While farming the potatoes, his real ambition was to raise cattle and added to his little herd whenever possible. As time passed, they quit raising potatoes and just produced feed for the cattle such as corn and alfalfa. Most farmers repaired their equipment in the winter, but Tub worked all year round, raising feed in the summer and feeding it in the winter. The machinery suffered from the constant usage. Tub used baling wire to patch the equipment. They called him, the "haywire king."

It was about 1953 when Tub and Wanda drove to Quincy one day to get parts for a Case tractor. When they came to the east city limits, they saw a large crowd gathered at the bridge. Wondering what was happening, they stopped, got out of the pickup and asked what the excitement was all about. It seems they came by just in time for the first water delivery in the new Bureau canal. They joined the historic moment, watching as the water approached pushing a huge pile of tumbleweeds past the cheering group of people. They had been hearing and reading about the dam construction and the water deliveries for years, but were busy with their own farming and it was a pleasant surprise to see it arrive accidently.

When water from the Grand Coulee Dam arrived, the Hansens didn't sign up for water. They kept their ground out of the project because they still had their own water delivery system. By then they also had a farm on O'Sullivan that was set up with wells and they excluded that too. They only had one unit that used Bureau of Reclamation water. Therefore, the project didn't make a big impact on their farming. It did take away some of the range land. Their cattle used to run on ground from Mae Valley up to almost Ephrata. Some of the ground in that area came under water delivery and was lost for grazing.

They had 30,000 acres deeded and leased ground in the Sand Dune area between O'Sullivan Lake and the home place in Mae Valley. When the government started to build O'Sullivan Dam, the Bureau of Reclamation condemned and appropriated all of the ground and generously paid them $8 an acre for it. The agency was very nice about it though, they rented it back to them for more than that sum over the next few years.

The water availability not only hadn't helped the Hansens, it obviously was detrimental in some ways. Tub and Wanda could see the need for the water and it was turning the area into the most productive farming area in the nation They didn't turn against the new era of agriculture in the county, but grew along with it.

One day, Tub, Bud Saunders, Harold Schwab, Bob Goodwin and one other Moses Lake man had been to a rodeo out of town because there wasn't any at home. They sat under the tree in Tub's yard discussing the poor show they had witnessed. They concluded they could put on a better show than that. They decided to form a rodeo group and assessed all the members enough to put a rodeo on. It was a struggle for a few years to make ends meet financially. It was a labor of love or they would never have succeeded. Those five people were responsible for starting what is now a successful, well known rodeo in the northwest.

Tub was elected to the Agricultural Stabilization and Conservation Service Board for Grant County. It was a three-man board made up of farmers elected by farmers to oversee programs made available by the federal government. He was very committed to doing a good and fair job for farmers in Grant County and represented them with determination and vigor. His board had a running battle with the Washington State Soil Conservation Service over unwarranted turn downs of conservation practices. The confrontation lasted a couple of years before Tub and fellow board members, Paul Holman and Bob Holloway finally won their point with Tub's able leadership. But only after requesting and getting help from U.S. Representative Catherine May, did they succeed.

The Washington State legislative elections of 1972 saw Tub winning a spot in the House of Representatives for the 13th District. He was seated in 1973 and served there until 1978, when he was elected to the Washington State Senate. Wanda was quite apprehensive when they went to Olympia. She had never been away from home without any of her children. She was used to being on the ranch and this was a whole new world, everything was new to her and they had heard politics were so dirty. She was afraid if Tub made the wrong vote, he might end up with his throat cut.

The whole family came to Olympia when Tub was sworn in the first time. The next morning the family went home and he went into session. Wanda was in the rented house alone when she thought to herself, "I can just sit here in the house and get lonesome or I can orient myself and do something." She knew how to get from the house to the Capitol grounds. She went to the gallery and watched the proceedings and felt much better with Tub in sight down below on the floor. When Tub saw a familiar face, he felt more at ease because everything was new to him as well. The security people in the gallery got to know her and would greet her each day.

"It was good to have somebody to say hello to," she said. At first she couldn't understand a thing they were doing or saying. "It was worse than going to a livestock auction."

Gradually things started making sense and she could grasp the procedures. Tub was "really and truly" busy and asked Wanda to follow a few of his bills; he had three the first year. He told her he was too busy putting out fires on bills that were no good to enter many of his own. She didn't know at first how to go about following the bills through the Senate, but she learned, and by that time she was "hooked". The legislative process had become addictive. From then on during the sessions, she spent nearly all of her time in the gallery of the House or Senate.

They found that everyone was nice to them. Sometimes they wondered who were really friends and who just wanted something because of his vote on a particular issue or something else Tub could help them achieve. But all in all, they were there at a good time because there was a lot of cooperation between both parties and each would actually consult one another on issues regardless of party affiliation. Wanda was impressed with the system as a whole, people of different professions and backgrounds, all working together for the common good.

Al Bauer was a Senator from Goldendale. He had a ranch near town that had run out of water and was in dire trouble if he couldn't get a good well. Tub had the ability to witch water. He helped many people over the years locate the best spot to drill for irrigation wells. Tub offered to go there and find a drilling site.

Tub called Bob Holloway and asked if he had time to fly him to Goldendale. They left Ephrata in a stiff wind and landed at the Goldendale airport where Al was waiting to transport them to the ranch. Tub spent about 30 minutes covering the area.

"Drill it right here," Tub said, as he marked a spot. Bob used a metal clothes hanger to bob for depth, a trick he learned from Arnold Greenwalt of Quincy. He got two depth readings for water. Al took Tub and Bob back to the airport where they took off for home. The whole thing only took a few hours.

A week later Tub was back in Olympia when he called Bob and asked if he remembered the depth the bobbing had indicated. Bob wasn't certain, then Tub asked him if he would fly back down and check the spot again and call him back. When Bob returned from Goldendale, he called Tub to say his lower reading was somewhat lower than the depth of the drilling. The new well was already a little deeper than the upper reading, but still lacked 20 or 30 feet to the second level. "Keep going, don't change the spot," Tub told Al when he called him.

Tub worried half the night wondering if he had done the right thing. What if they didn't get any water? Al's ranch depended upon it. A day later Al called Tub in Olympia. Wanda was across the room, but she could hear Al practically screaming. Tub was holding the phone way out from his ear because Al's voice was booming in excitement saying they had hit water, not only that but it was an artesian and the water had filled all of his old dry lines and was coming out the top. He could use all of his old irrigation system. Al stated at

Tub's memorial service years later that Tub had saved his ranch and it wasn't an exaggeration. Al still serves in the Senate at this date in 1999.

Governor Gardner, in 1989, wanted to do away with the Wildlife Executive Board and appoint the director himself. The method at the time, and still is, the governor appoints the board from around the state and they choose the director. The Democrats, at Gardner's direction, had a bill they were trying to pass to achieve that end. Tub didn't believe in it and thought it was strictly a political ploy. He was the only Democrat in both houses to vote against the bill. They could pass the bill if Tub would vote yes. They put all kinds of pressure on Tub and gave him a bad time.

Wanda came into the room at the Capitol one day where Tub was talking with someone as she passed. Tub took her arm and said, "Lets go up to my office, I want to talk to you."

When in the office, it was the pressure to vote for the bill he wanted to talk about. The Governor's office had been phoning people in his home district to call and put additional pressure on Tub and it was bothering him.

"I'm not going to do it, it isn't right, and I'm not going to do it," he said to Wanda.

Afterward when Wanda went to the gallery, Becky Bogart, a lady who was the Governor's liaison, called Wanda out of the gallery to talk. When she went out Becky said, "Tub has got himself boxed now and can't change his vote, he's got to find a way out, is there anybody we can call and have them call him and give him a way out so he can change his vote?"

Wanda said, "I just talked to him five minutes ago and you are not going to get him to change his vote, you've got to find it somewhere else."

When the vote was taken, Tub voted "NO" and the bill went under. He really and truly was a man of his convictions. "I'm so proud of that," Wanda said later.

Wanda described their lives in segments. The first 20 years they were growing up. The next 30 years they were building a cattle ranch and raising their family of four children, Tom, Jerry, Kay and Penny.

The following 20 years was an entirely different life in the legislature. Tub's goal was to spend 20 years in tbe legislature, but he passed away one year shy of that goal. As she had for a lifetime, his wife and partner, Wanda stepped in and ably filled out the remaining period to achieve his commitment to himself and those at home whom he served.

For the existing segment, Wanda is building a life for herself completely separate from all that has gone before. She said, "We were very lucky." She now spends part of her winters in southern Nevada with her son Tom and the summers in Moses Lake.

Tub's memorial service was one of the largest ever held in the area. He passed away in late December of 1991. The crowd was made up of cowboys, farmers, city businessmen, senators and representatives from all over the state, and all kinds in between. Everyone knew it was the passing of a rare man, a humble man, a man that took what God had given him, and along with a wife who was his equal, made it from a hungry rodeo performer to a respected member of the Washington State Senate in Olympia. He and Wanda were just as much themselves with a ranch hand as they were with the Governor. His grammar may have lacked polish, but not what he said. When he spoke, everyone paid attention.

While Wanda was serving in the Senate after Tub's passing, the state appropriated funding for the Olmsted State Park between Ellensburg and Kittitas. The budget included funding for a sun-dial memorial to Tub.

Wanda made an amazing transition for a country girl out of Idaho. She helped develop a successful ranch operation while raising a family under extreme financial conditions. She served in the Washington State Senate, while being a charming, elegant lady all through the process. Like many others in the Basin, Tub and Wanda Hansen were true pioneers and nation builders, "really and truly."

Wanda Hansen supplied both written and oral information in the preceding history

Tub and Wanda, 1946

Tub and Wanda, 1989

Haidi and Shigeko Hirai

I'm Married Now and Can't Afford Coffee

Now this was an unlikely combination. Haidi had a background of farming and was strictly a country boy. After high school, Shigeko trained, worked and lived in the big city and knew nothing about agriculture. Yet, because of a similar ethnic background and a world war, the fragile web of circumstance wove a pattern to bring them together in the newly watered desert of Central Washington.

Originally, Haidi was from California where his family operated a nursery. When war with Japan broke out, the family was evacuated by the U.S. Government, all their property was taken with no compensation and they were sent to a camp near Minidoka, Idaho. Even U.S. born citizens, if they were 25% Japanese or more were included. There was a roundup of Japanese people in South America as well as North America.

After release from internment in Idaho, the Hirai family migrated to South Dakota. It was there they heard irrigation water would be available in the Columbia Basin of Washington where people could start a new life. It was 1953 when they arrived in the sagebrush country south of Quincy.

Shigeko Iseri graduated from Chimacum High School on the Olympic peninsula in Washington State in 1935. Her parents were dairy farmers at the time of the attack on Pearl Harbor. There were seven children in the family. They owned and lived on 130 acres with the dairy cattle. Her father had come to the U.S. when he was 18 years old. They were good, law abiding, productive citizens. Because of their ethnic origin, Japanese, they had to sell the cows and the Government confiscated the dairy and land. They took everything the family had, including their bank account and paid them nothing. All they could take with them was the clothes they could pack.

Shigeko was in nurse's training at the time and was given less than a day to pack and leave the school, otherwise she would be sent to a camp apart from her family. Not wanting to be separated, she went home where within a month they were shipped to Tulle Lake, California for internment for the duration of the war.

The family was put in Block 18 within the internment camp at Tulle Lake. They were located next to the police headquarters according to Shigeko. They worked while in the camp and received wages of 14 to $19 a month. The house was a one room structure for the entire family of nine.

It was 1990 before the families received any compensation from the U.S. Government for the businesses and homes which were taken away from them in 1941. Each person received $20,000 for compensation after almost 50 years. It was enough to buy Shigeko a car in 1991.

It was the late 1950s that Shigeko had an acquaintance in the hospital where they worked who knew some Japanese farmers at Quincy. They found Shigeko's parents and the elder Hirais had known each other in the past. So it came to pass Shigeko and her friend took a train to Quincy one weekend where she was introduced to the Japanese young people of the area. Most of the young people were young men, there were very few girls in the area. She met the Maruokas, Tsukamakis, Higashiyamas, Manos, Kikuchis, Gotos, and of course the Hirais, including Haidi.

It was three years before she heard from Haidi again. He telephoned one day and said "I have a message from your girlfriend." She replied, "that was fine." He said "Are you going to be home?" She answered "Yes, I'll be home but I'll be on call." He said he was coming over. On the way he stopped at the telephone booth at Ellensburg, pulled out a little black book from his wallet and called to see if she was still there. When he arrived in Issaquah, he phoned again, wanting to make certain he wasn't wasting a trip if she happened to be called to work. He arrived in Seattle and was running low on fuel, stopped to fill the tank at a station and discovered his wallet was gone.

He proceeded to her apartment and said, "We've got to go back, I lost my wallet somewhere." They went to Issaquah, no one had found it or turned it. Then they headed for Ellensburg where he found it still sitting in the telephone booth where he had laid it down when removing his little black book. That was their first date, it probably showed he really did need someone to watch over him.

Haidi never called from the Hirai residence, evidently not wanting the family to know he was interested in someone. When Shigeko would come to Quincy to visit her friend, who was married, her friend took off her wedding ring so no one would know which was the girl friend. They kept the family and friends in suspense trying to guess which girl Haidi was courting.

It was a few years later when Haidi came by the way of Seattle as he was driving to California. Shigeko wasn't home so he got her room key from her friend Sonia who lived in the same building. He left an engagement ring in the room by the telephone and took off. After he left, he phoned back to Sonia and asked if she would check the door, he said he couldn't remember if he locked it or not. In other words she didn't receive the ring directly from him. He later wrote saying, "Don't make a fuss over it."

Before they were married, Haidi showed up at Earl and Ruby Holloways while Holloway's daughter and her fiancé were visiting. He asked the daughter, Betty, and her friend Ray if they would help him sort his papers for income tax. When they said yes, he ran back home and brought down shoe boxes full of receipts, dumped them on the living room floor and all three of them started sorting. It was real early a few mornings later when Haidi came over while Betty was still sleeping and lit firecrackers underneath her bedroom window.

Haidi and Shigeko were married in December of 1960 at Seattle, Washington with about half of the town of Quincy in attendance at the reception. There was singing, dancing and story telling to the delight of all who were present.

Haidi used to stop at the neighbors for coffee. He once told Ruby Holloway as they were at the kitchen table sipping coffee, "Ruby, I'm going to put you in my will and furnish you with coffee the rest of your days." A short while after getting married, he stopped by and said, "Well Ruby, I have to go back on that, I'm married now and can't afford to give you that coffee."

When Shigeko moved to the farm, she had a hard time finding where she lived any time she went anywhere because every thing looked the same, sagebrush, dirt roads with no street signs, no landmarks to go by like the hills of Seattle. When she went to the field to get potatoes for dinner, she didn't know if she was in their own field or the neighbors.

While Haidi farmed, Shigeko worked at a doctor's office in Quincy after moving to the Columbia Basin. The doctor flew a small plane and always wanted to take her up, much to her distress, as she was leery of small airplanes. She must have been psychic, because the doctor and his son were killed in his plane a few years later going fishing in Alaska.

Farmers being rather independent, didn't like being told what to do and how to do it. Haidi was no exception. One summer after wheat harvest, he wanted to put in a crop for the following year which needed to be planted early in the fall. The new crop required cleaning the field so the seed could be planted properly. The best and quickest way was to burn the stubble. The fire not only killed weeds without chemicals, but destroyed volunteer wheat as well.

Shigeko was ironing one day when all of a sudden she heard sirens and a commotion. When she opened the door there were fire trucks pulling in all over the yard and firemen running to the door shouting "where is the fire." She was almost speechless and finally was able to say she knew nothing about a fire. It was then they saw the smoke coming from a nearby field where Haidi was diligently burning his field but with no permit so when the fire station was called they had no knowledge of a controlled burn. I think he called them the next time.

Shigeko said she never did become a farmer or understand what was happening on the farm, that was Haidi's area. One day Haidi asked her to go one direction on the ditch setting water with syphon tubes. He would go one direction while she went the other. She had never set a syphon tube before and had no idea how to proceed. She thought, "What should I do, stand on my head with my butt in the air and suck water out of the ditch or what?," but the tube was too large for that; so much for learning modern irrigation. Haidi asked her to take care of the farm books and pay the bills. She said "I never saw such bills, I don't know how to pay these big bills."

Haidi had two younger brothers, Tom and Jim along with four sisters, Nancy, Betty, Marianne and Patty. The girls got married and left the area. Tom and Jim became successful in farming as well as businessmen. Jim was killed in an auto accident in 1980. Mt. St. Helen's ash kicked up by a passing car made the visibility so dense that Jim and a neighbor, Chet Ashy, collided head-on resulting in Jim's death.

In 1962 Haidi and Shigeko went to Japan and brought home two adopted children, Kenny and Shirley. Shigeko was 45 years old and it was a dramatic change for both her and Haidi from the home life to which they were accustomed. It seems parenting was harder then the farming.

All the Hirais were very good farmers. Haidi was also quite active in the community and worked with Washington State University in many agricultural projects. He was a popular resident of the Quincy Valley and served on The Washington State Potato Commission. After conferring with Dr. Faulkner from WSU, Haidi decided Chewelah would be a good place to raise certified seed potatoes. It was the fall of 1969 he started building on the land he acquired and by December Shigeko joined him at their new seed farm in Eastern Washington, not far from the Idaho border.

Everyone who knew Haidi thought of him as thoroughly Americanized in nearly every way. There was an underlying old country culture that maybe he didn't even realize existed in him. It was March of 1970 and the loan he had sought all winter still hadn't come through. Haidi had always been successful at everything he tried, but this loan to get the crop in was essential to the new enterprise.

On the 17th of March Haidi went out by himself without a word to anyone and committed suicide thinking he had failed and he couldn't accept that. He was found later that day. It was a sad day for everyone who knew him. His zest for living and knowledge of farming were lost forever. The irony of the whole episode was the very day after he had left, the bank tried to inform him the loan had finally been approved.

Brother Tom took time from his operation in Quincy to go ahead with the seed farm and manage it for Shigeko for a few years until they rented the land; and then in 1985, Shigeko auctioned it off and quit. Shigeko still resides in Chewelah and visits old friends in Quincy as she passes through to Seattle to see relatives.

The Hirais are a typical example of the American Japanese families' treatment during the war with Japan between 1941 and 1944. Even though they lost all their property and money to the U.S. government, there seems to be no resentment toward their treatment. They all took it with dignity and started all over from scratch when released. No other ethnic group, so treated, that wouldn't have screamed to high heaven about inequality and discrimination. They are living proof of the opportunity available to citizens of this country to make better lives through hard work and sacrifice. They are examples of the great American Dream which they fulfilled in the new Columbia Basin Irrigation Project.

Paul and Virginia Hirai

Choices

It was June, 1942, the family of 15-year-old Paul Hirai was told to move from their home. The orders came from the United States Government. They were given no choice but to evacuate quickly. A few months earlier on December 7, 1941, Japanese military forces had attacked Pearl Harbor and war was declared against Japan. All United States citizens of Japanese ancestry went through a dramatic change in their lives. The authorities gave them only a few days, forcing a quick sale of their equipment and the growing crops on the rented 40 acre farm. The United States military was paying good prices for produce during the war. When the crop from their acreage was harvested 30 days later, the buyer received a net profit of $40,000 after paying the Hirai family $500 for everything.

Paul was born in Toppenish, Washington. His father started farming there about 1930. Paul was seven years old when his father passed away. His mother, older brothers and sister continued the truck garden farming operation on rented ground.

They sent Paul and his family to a Portland, Oregon assembly center. No housing was available in Portland so they converted the animal stables at the fairgrounds into temporary housing. They constructed additional tar paper shacks to hold everyone. Others from Yakima and the Portland vicinity joined them. Communal kitchens, mess halls and bathhouses were set up to accommodate their needs.

A year later the authorities decided to move them inland from the coastal area. Most of the people were sent to Minnedoka, Idaho and Heart Mountain, Wyoming. A farmer they had known in Yakima Valley, had moved to Nyssa, Oregon, and developed new land. He requested the authorities to allow nearly 300 people, including the Hirai family, to come to Nyssa on a work project. Labor was in short supply for raising sugar beets. That farmer, Sid Flanagan, later moved to Quincy to farm and served in the Washington State legislature for many years.

Paul and his family thinned beets the summer of 1942 and stayed in a tent camp during the following winter. Sid Flanagan led an effort in the community to build housing in labor camps around the area. Paul's family opted to stay on the farm of Ken and Roland Goulet in a converted outbuilding. While the rest of the family worked for Flanagan and the Goulets, Paul at 16 years of age had a choice. He could attend school like any other normal teenager or he could help his family to regain what they had lost. Paul chose to rent a 16 acre field and put in an onion crop. Paul's brothers worked for the farmers in the daytime and helped Paul in the evenings. He kept experimenting and increasing the acres until he was up to 80 acres of onions. During this period he was also attending high school at Adrian. Although his mother, sister and two brothers were paid a dollar a day each, the family managed to save money. Paul borrowed machinery from the Goulets and other neighbors. As they saved money from their labor and Paul's crops, they bought their own equipment one piece at a time.

Toward the end of the war and a few years afterward, vehicles and machinery were hard to come by. In their spare time the brothers went as far as Chicago to buy used cars and into Washington State locating used equipment and transporting it back to Nyssa. They either resold for a good profit or filled out their own line of equipment. Working the farm gave them an advantage with the amount of gas and tire coupons they received. They could travel where others could not.

Paul's landlord, an automobile dealer, owned several hundred acres farmed by the Goulet brothers. When the Goulets moved to the Columbia Basin, the landlord asked the Hirai brothers to farm all of his ground. He helped finance additional machinery they would need. Besides farming, he wanted them to move his cattle in Burns, Oregon, to the farm and feed them during the winters. Over a few years time, they were farming several hundred acres. They were feeding three to four hundred head of cattle. To top it off, they took on the job of feeding hogs that had been running wild on the Burns, Oregon ranch and moved to Nyssa.

Paul and Jim Cables met in the National Guard and become best friends. In 1951, Jim wanted to visit Fort Benning, Georgia. Since Jim's father was a new car dealer, he took a car off the show room floor and away they went. They drove to Fort Benning where they attended officer candidate school and came out as second lieutenants. They covered the southern states on the weekends in style with Jim's new automobile.

Virginia (Ginny), was born in Seattle in 1932. She was one of six children. When she was eight years old, her parents decided they wanted their younger children to receive Japanese schooling and moved back to Japan in 1940. Knowing certain things were scarce in Japan, their parents brought extra everyday items from the States. Boxes upon boxes were loaded on the ship. The boxes were filled with household supplies and oversize garments and shoes for the younger children so they would have the right size for future needs. It was a relief when unloading from the ship in Japan that Ginny's cousin, being very influential, passed their belongings through customs without having to sort through every item.

Her parents had originally been from the island of Shikoku. They moved to a beautiful rural fishing village on an inland sea by the name of Ehime. It was a great place for the younger children. Everyday they enjoyed clamming and fishing on the wharf two blocks from their home. Ginny's oldest sister was married and remained in the U.S. The second oldest went with them for a vacation and returned to the States. Her third sister, due to health problems and not wanting to live in Japan, was brought back to the States by her mother early in 1941. Her mother returned to Japan to join her husband and younger children shortly before war broke out. The family was separated without contact during the war in the Pacific.

The stay in Japan during the war was not pleasant. There was a shortage of food among other things. By comparison, the living standards were not what they were in the U.S. The family, although Japanese, were still American citizens and were under constant surveillance. Periodically, the authorities questioned her father and mother. Ginny was too young to know how closely they were observed. The extra clothes, shoes and other articles they had brought from the U.S. turned out to be of great help. The children were bilingual and had attended a Catholic missionary school. Schooling was not hard, they could already speak and write English and Japanese and had no problem fitting in.

After the war ended, her brother returned to the U.S. in 1947. When the Korean war started, Ginny became apprehensive. She had two choices, stay in Japan, or return to the U.S. as a citizen. She didn't want to be stuck in Japan the rest of her life. In 1950, she also returned to the States. That left her parents and younger sister in Japan until her father died in 1952. Shortly afterward, her mother and sister came back to this country for good.

Ginny finished her high school degree at Edison Technical School in Seattle. She attended Seattle University for a couple of quarters and was trained in bookkeeping. One of Ginny's sisters was married and lived in Nyssa where she travelled to visit her ocasionally. While there she met her brother-in-law's brother, Paul. When Paul went to his two-week National guard training each summer in Seattle, he stopped to see Ginny and her mother.

Another farmer friend of the Hirai family came to the Basin from Ontario, Oregon in 1952. Paul helped Harry Masto move to Moses Lake but resisted Harry's efforts for the Hirais to come with them. Paul and Virginia were married in 1955. They continued farming in the Nyssa area and by 1959 were up to 1,000 acres in crop. The acreage was scattered around causing lost time in moving equipment and extra miles driving back and forth. Harry, on his yearly visits, kept encouraging Paul to move his operations to Moses Lake and become his partner. Again they had a choice to make, to stay where they were successful, or go to a new area and take the chance of greater success or perhaps failure. Paul and Virginia accepted Harry's offer in 1960. They became partners and planted 300 acres of potatoes the first year. The prices the first few years were at disaster levels of seven dollars a ton. They chose to stay with the potato program and by 1965 were raising 3,000 acres.

Pronto Foods built a new potato processing plant the fall of 1965. Harry was a major stockholder, giving them the opportunity to raise more potatoes. He and Paul were the main suppliers for the plant for a few years. The plant had financial problems the first three years and most farmers were not willing to risk contracting to them. Paul and Harry ran the off grade potatoes through the processing plant and the rest through two fresh produce sheds of their own, one in Moses Lake and one in Othello. Paul was not only raising potatoes for the plant, but loading the crop out of storage during the winter months for delivery to the plant. He was also responsible for disposing of the plant waste water until 1972 when Carnation Company purchased Pronto Foods.

The potato sheds continued marketing their crop until 1975 when they chose to convert to raising and packing onions. By the late 1970s, they were farming almost 5000 acres of irrigated crops northwest of Moses Lake near Wheeler. Most of the acres were in wheat. It was used in rotation with dry beans, potatoes or onions. Harry retired in 1984 and Paul took over the operation through 1993 when he sold and retired from farming and the produce business.

The most enjoyable part of their life, for Paul and Ginny, has been raising three daughters and seeing their grandchildren; of which they are very proud. The daughters attended Moses Lake High School. One went to Kinman Business School in Spokane. The other two sisters attended Eastern Washington University. The oldest is now

Director of Human Services for the Moses Lake school district. The second daughter is married and operates her own escrow business in Moses Lake. The third daughter is married and lives in Salt Lake City, Utah.

Paul became interested in community service while still in Nyssa as a member of the Junior Chamber of Commerce. He enjoyed working on projects such as Easter egg hunts for kids and selling Christmas trees. Paul joined the Moses Lake Rotary Club in 1972 and has worked on many projects with them over the years and is chairman of the World Service Community Project. The scope of the project is using money they have collected to help in foreign third world countries. They have built water systems in the Philippines, helped Russians with food and book programs and built several homes in the Fiji Islands. A continuing project in Bogata, Columbia has raised $30,000 to help cleft-mouth children. These programs have been supported by local involvement and matching funds from Rotary International and help from a Rotary Club in Japan.

Starting in 1998, the Moses Lake Rotarians collected $70,000 from clubs, districts and Rotary International. They purchased and distributed 13,000 textbooks to Guatemala school children. Paul is very soft spoken, articulate and calm. Nevertheless, when speaking about handing out school books to children who have never had any, his eyes sparkle with excitement and obvious pleasure.

Early in the year 2000, Paul and Ginny travelled to Guatemala. Joining them were John Townsend, Mike Townsend and his two daughters, LaDel Yada and Patsy Lee, all from Moses Lake. Rotary Club members from Ohio, two from North Carolina, one from Atlanta, Georgia and Ron Ritz from Leavenworth also came with them. They all had the satisfaction of travelling to the primitive highlands of Guatemala. There they personally handed out 20,000 text books bought with $102,000 raised during the previous year. The group also took with them, writing pens, pencils, basketballs and soccer balls to give to those who have no way to buy such items that U.S. children take for granted.

Two brothers, Jeff and Joe, from Ohio, coordinate the project in Guatemala for Rotary. They live in the country and make all the contacts with schools, officials, etc. Both brothers had high paying corporate level jobs at home in the States, but quit to do more in life than make money and be comfortable. They wanted to do something that would make a difference for humanity, not too much different from Paul and Ginny's own goals.

Jeff originally went to visit an uncle who left the priesthood, married and was living in Guatemala. Jeff decided to teach English in the area temporally. That is when he noticed the need of textbooks in the secondary schools. Jeff resigned his job with Proctor-Gamble. He talked his brother Joe into quitting IBM and move to Guatemala to help in his quest to assist the schools and the people.

Joe spends time in the U.S. trying to raise funds with sponsors such as IAM Dog Food and Steel Fabricators. They utilize all the money collected by Rotary for textbooks. Not a penny is used toward administration or the brothers livelihood. While working out of their uncle's home, they met and started working with another Catholic, Father Bernie, who is spending his life in the same area.

Guatemala subsidizes schools up to the sixth grade. Probably for lack of funding, they ignore the secondary level of schooling, so no textbooks. That is where Paul's Rotary program steps in to help fill the void. The trip in 2000 was the second for Paul and Ginny. On the first visit, Ginny took pictures of the children receiving the supplies and small gifts. The second trip, she took the photos back knowing they would see some of the same students. The students had never seen photos of themselves before and became quite excited and filled with wonder as Ginny exhibited them. To Ginny, it was pure enjoyment to see the facial expressions of happiness and awe as they gathered around her clutching her hands. Paul said the reason he works so hard on this particular project is because it's not a give away program. It is set up as a rental program and will be self supporting after four or five years. The project starts with the brothers consulting with a given school. They call a meeting at the school, the parents have to be there because they will be the ones ultimately paying for the books. The community leaders also have to be present because they do not trust anyone coming into their area. If they cannot get the approval of all three entities, they won't go there. If the school, the parents and the officials accept the program, they are charged very small rental fees for the books. They make payments once or twice a year to the nonprofit organization and deposited in a U.S. bank. When the textbooks wear out, they will distribute new ones, the old ones going to other schools that haven't any. There are 600 secondary schools and the group has covered 51 of them in the first three years. The goal is to reach the other 550. They ordered and printed the books within Guatemala.

Paul said the generosity of the Moses Lake people on the trip was overwhelming. At one location it took great effort for 4-wheel drive jeeps to make a 40 minute drive from the road up to the school. It is an area where it rains six months out of the year. The teachers have to walk three and a half hours every day just getting to and from the road to the remote school. The school was closed for a long time before two teachers were found who were dedicated enough to take the job. The group asked the teachers what they really needed the most. The teachers weren't thinking of themselves. The first item was a first aid kit for the kids. The next were boots and umbrellas for themselves for the trek in the rain each day. The salary they receive wasn't enough to afford such luxury. The Moses Lake people decided to take a collection and give it to the teachers at a public ceremony. They were warned that if even that small amount of money were known to be received, the teachers could be killed to obtain it. They gave the money to them privately, along with a ride back to the road at the end of the day.

Father Bernie is a Benedictine Monk and has been working in that area for 20 years working out of a monastery. He led the group to a site in a high location where one can see miles and miles across country. He wanted to move a radio station to that spot and broadcast to the surrounding area. They would hear it across the border into Honduras. Father Bernie didn't own the ground and wanted to buy it but didn't have the funds. When asked how much ground and what the cost would be, he pointed out a fence that surrounded about three acres. The cost would be $600. Paul and John Townsend dug into their pockets on the spot and provided the necessary money for the purchase and fulfilment of his dream.

They visited 10 different schools on their tour distributing textbooks. Not used to seeing foreigners, at each school they stopped, the parents and local authorities would gather to meet them. The American brothers, Jeff and Joe Berninger, have gained the respect of the school administrators. They were told, "The government makes us promises, but they don't keep them, not one word. Your people are the only ones who tell us you will do something and you do it."

Paul and Ginny have been hosts for Japanese exchange students at their home and farms. Paul has led tours and made arrangements for accommodations for students for more than 30 years. In his passion for furthering education, he feels he has been fortunate to serve Big Bend Community College, since 1982, on the Board of Trustees. He feels it has been a highlight of his life. He has met good people, enjoyed serving and hope he has made a difference by participating. In 1999 they appointed him as a member of BBCC Foundation Board which he finds very gratifying. Paul and Ginny both say they think the nicest people they have met in life have been associated with the college and Rotary. Paul received the following in November of 1999:

Mr. Paul Hirai
Moses Lake, WA 98837

Dear Mr. Hirai:

The Order of the Sacred Treasure, Gold Rays with Rosette, was conferred upon you today by H.M. the Emperor of Japan. This decoration is in recognition of your outstanding contribution to the promotion of receiving Japanese agricultural trainees, and to the promotion of understanding between Japan and the United States of America.

The conveyance of the Order to you will take place on a mutually convenient date at my residence.

I wish to take this opportunity to express my most sincere congratulations to you for the conferment of my Government's decoration.

Sincerely yours,
S. Sato
For, Yoshio Nomoto
Consul-General of Japan

The following story appeared in December 1999 in the Columbia Basin Herald:

Emperor Decorates Moses Lake Japanese American

Paul Hirai, former Chairman of the Big Bend Community College(BBCC) Foundation and a long time farmer in Moses Lake Washington, was decorated by the Emperor of Japan on November 3, 1999. The decoration ceremony for Mr. Hirai was held at BBCC in Moses Lake on December 15th, as many as 150 people attended the ceremony and Consul General Nomoto presented this order. Mr. Hirai received the "Order of the Sacred Treasure, Gold Rays with Rosette" for his 33 year role serving as a bridge between Japan and Moses Lake, Washington, through various human and cultural exchange programs such as the Japanese Agriculture Training Program and the sister city program.

Mr. Hirai, then a leader in the Japanese community in the Moses Lake area, successfully persuaded the local Community College into hosting the Japanese Agricultural Training Program in 1966. The two-year program, which provides the potential Japanese farmers with the opportunity to learn about American style farming, brought 4,000 Japanese trainees to mainly Washington, Oregon and California over the past 33 years. Mr Hirai was also the key promoter in establishing a sister city relationship between Moses Lake and Yonezawa, Japan in 1981. The affiliation resulted in exchanges of numerous friendship delegations between the two cities.

Besides the two-year program, there is a one-year and Extension Agent program which Paul has supported. He has been active with the sister city program for 20 years where they send 50 local students, plus chaperons to Japan every year. In 1985, Paul and Ginny went to Japan on the 20th anniversary of the Agriculture program. They met and shook hands with the "Crown Prince" of Japan at a reception. The same Crown Prince, is now the present Emperor of Japan who gave him his award. Paul and Ginny feel very good about the medal. It has more significance to them since it comes from someone whom they have actually met.

Another project Paul has worked on for many years, is a Japanese Garden Committee. They are attempting to build a Japanese park in Moses Lake with the aid of Washington State and the city. Construction may begin the fall of 2000 in the three ponds area. Included would be a Japanese style bridge within the park.

Besides helping and supporting Paul with his projects, Ginny is active in an investment group of local women that formed in 1982. At first they wanted to learn more about finances and the market. Speakers and mini-seminars were held to educate themselves about economy and finance. While learning, they have invested in the stock market and not only made some money, but had fun doing it. They limit the number of members to 20 and have a waiting list of those wanting to join.

She is a board member of the Columbia Basin Allied Arts which promotes various programs including musicals and plays. Each year they sponsor four major events for adults and four for children's entertainment. All of the community activities require funds to operate and she spends many hours volunteering for events like garden tours to raise money to keep Allied Arts operating. Besides the other activities, she has worked as bookkeeper for Dress Ranch and Home in Moses Lake for many years.

After Paul's retirement, he and Ginny could do as many others have done, relax and play golf, travel at their leisure and let the rest of the world go by and all their time be their own. After all, they had earned the right to do just that, or they could choose a different path. They chose to work for the advancement of education and the arts.

The service work they have done and anticipate doing, is what Paul calls "pay back time to the community." Paul said, "I love the people, I love the climate, it's a wonderful place. We are doing what we want to do and at the place we want to do it."

Paul and Ginny epitomize the American dream. No outcry came from the Hirai family that they had been treated unfairly. They accepted the situation that existed and made the choice to improve it. Paul and Ginny, through desire, persistence, planning, frugality and plain old fashioned work and making the right choices, overcame the obstacles and turned them into opportunities. They took the responsibility of trying to enrich not only their local community in education and the Arts, but around the world as well.

Ginny and Guatemala school children

The Rotary group with some of the school children

Paul and Gunny at the award ceremony

Earl and Ruby Holloway

A Home of Their Own

Life was not meant to be easy. Most of the early settlers took that for granted. Life is not fair. Everyone is not guaranteed equal education, freedom from poverty or happiness. But some break the mold and achieve their dreams, gain knowledge, help others and find happiness in doing it.

It is quite interesting how mobile people were that were born in the early 1900s. Even though the transportation was more primitive and slow, many of the people who came to the Columbia Basin saw a lot of country before arriving at their eventual home.

Earl and Ruby were an example of the search for a better life for themselves and their children. The majority of people of that time lived in rural settings and struggled to maintain food and clothing. Hardship was the norm rather than random. They were by and large, quite poor, but they didn't know it, it was accepted that life was not easy.

Ruby was born at home in a small town in Missouri on Christmas Eve 1912. She had four sisters and five brothers. By the time she started school, the family moved to Wyoming, back to Missouri, to Colorado and finally to Twin Falls, Idaho, in 1917. The last move was on board a train.

Twin Falls at the time was a "dusty old hole" with a small population and unpaved streets. Her father worked northeast of town for three years, than moved to Greenwood, Idaho, where they bought a small farm. It wasn't long before moving to Dixon, Idaho, where they resided until Ruby, a friend and two other girls went to Twin Falls to work. The first job was at the hospital serving and cleaning in the lunch room. After that she worked sorting beans in a local warehouse. It was hard work, but still better than weeding in the fields.

Earl was born February 2, 1907 in Marion County Kansas. He was the youngest of 10 brothers and sisters along with two cousins raised with the family. Earl's father covered a lot of territory. While the children were still young, they left Kansas and travelled all the way to California. They passed through San Francisco in its earlier stages of growth. He took his family up the western coast and the northwest before stopping in Nez Perce, Idaho, their moving on to Kuna, Idaho, where they settled for a few years. Finally, they travelled to Twin Falls, the end of their journey. Earl helped drive the sheep wagon part of the time on the way.

Earl wanted to go to college, but due to financial conditions of the family, he was needed at home. If he got higher learning, it would have to be the kind you get on the job.

The Holloways were farmers without money to buy their own land. They sharecropped south of Twin Falls and Earl attended a small school at Hollister. His basketball team played other nearby towns including the larger town of Filer. The Filer team had a center, Web Jones, who Earl played against during high school. Web became a neighbor many years later after they both moved to the Columbia Basin.

Earl and Ruby met at a Twin Falls café where hamburgers sold for five cents apiece. They were married in Burley, Idaho, in 1930 and moved to Greenwood, east of Twin Falls where they lived in a small house and an old sheep wagon with up to nine other relatives. The quarters were crowded, but it was during the depression and work was almost nonexistent. There were other relatives that stayed with them off and on during the next few years, everyone depended on each other. Whatever one had, they all had a part. Earl was finally able to work for the canal company and his wages were what they all lived on.

The men rented farm ground, but the returns were almost non-existent. With nothing else to do, many times they would stay up all night playing rook, a card game, to pass the time. The small house was so crowded. Earl and Ruby moved into another house not far away. It wasn't long before the whole group moved in with them again. There were other moves to different places in the area, but each time, the clan followed as if they needed the strength of numbers to survive. The stove was used for cooking and a source of heating in the winter. The fuel consisted of trees which were cut and hauled out of the South Hills to keep from freezing during the winter months. Occasionally, a hog would be butchered for meat and the garden and field crops were the supply of food.

Their first child was Robert, born in 1931, the second and last was Betty, five years later. By then they were living by themselves in an eight by 20-foot shack in Hazelton. The shack had no bathroom or running water, but it was home.

It was 1936 when they decided to go to Oregon for the green pea harvest. Betty was only two months old when they made the first trek. It took two days to make the trip to Eastern Oregon. The first night lay over would usually occur in the Blue Mountains between Baker and Pendelton. They parked near the river. They built a camp fire for cooking and slept under the stars in blankets wrapped in a canvas tarp.

Earl worked in the pea fields loading trucks with a pitchfork. The loads would go to the stationary de-viner where the peas were shelled and boxed for the cannery. Needless to say, it took a lot of labor to harvest thousands of acres of crop. Earl also worked in the prune harvest in the same area.

Weston and Milton-Freewater were their homes during the summers they migrated to Oregon. They lived in a tent that was borrowed from Earl's cousin, Frank Holloway. They pitched it along with other laborers in outdoor camps, including Indians and their tepees. Betty, the youngest, was carried in a basket on the first trip.

They made this annual trip for several years and after harvest they got to know the surrounding country. One year they camped at the base of Steamboat Rock, south of where Grand Coulee Dam was being constructed. They marvelled at the suds piled up on the shoreline of Soap Lake as they came through the Central Basin area. One of Earl's brothers, Marion, had moved to East Wenatchee to work for an apple orchardist.

Just prior to 1940, Earl and Ruby started farming on their own near Twin Falls. Farm ground was expensive, they didn't have money to buy, so they rented for a few years trying to get ahead. The ground was sharecropped and too much went to the landlord, it was all work, risk and no gain.

After Pearl Harbor in 1941, they worked for a time in the shipyards in Portland. After the war, it was back to Twin Falls. They were able to buy an acre on the edge of town where they moved their small shack. Earl paid one dollar down on a Ferguson tractor and started doing custom work. He prepared the garden plots and leveled the yards of new homes being built. He got a reputation for doing good work and for the first time they were getting something of their own. They started a custom hay baling operation. Earl, Ruby and Robert worked on farms and ranches from central Idaho to northern Nevada baling alfalfa, straw and timothy hay.

The farm ground around Twin Falls was still priced out of their reach and after hearing about the Columbia Irrigation Project nearing completion, they became interested in trying to get a start in the desert they had passed through on their trips a few years before. Earl knew a man at Twin Falls by the name of Stromeyer who owned ground south of Quincy. They traded a house in Twin Falls straight across for the 80-acre farm unit. A short time later they acquired an adjoining 80 acre unit from a son-in-law of Stromeyer.

Earl, Ruby and Betty moved to the undeveloped ground in 1953. Robert was now married and stationed at McChord Air Force Base in Washington State. Ruby was happy to leave and start a new life for themselves. Earl was fulfilling his dream of owning their own farm. Betty was unhappy starting school in a new place among strangers, leaving her friends in Idaho behind.

The trek to their new home was reminiscent of the people leaving Oklahoma during the infamous "Dust Bowl" days. A ton and a half GMC truck loaded with all their belongings pulled a two wheeled, flat bed trailer. The trailer carried the little shack they had lived in several years earlier. They also had a small travel trailer and the Ferguson tractor with a one way two bottom plow.

They cleared sagebrush and leveled a corner of the ground to place their shack. There was no electrical power, no well, no inside plumbing. But they did have an old time outhouse. Only narrow dirt roads on the section lines were available to access the property.

The nearest neighbors were Leo and Teresa Healy who had lived and ran livestock for many years in the area. All the water was hauled from Healys. However, they did have hot water by putting a 50-gallon barrel on stilts and letting the sun do the heating. That fall Leo hurt his foot and Earl helped him with his water changing. In return, Leo had them move their shack in his yard so they could have water from the well and electricity.

Ruby had planned to buy calves to raise and sell. However, her plans went astray when she got the first one. His name was White Socks, and he became almost part of the family. Whenever anyone stopped, White Socks was there to be part of what was going on. It took a while to discover how the clothes hanging out to dry kept getting on the ground. White Socks was like a family dog, and Ruby wasn't going to raise calves like White Socks and then ship them to a feed lot and eventually the slaughter house.

That was when the real work started. They hired a crawler tractor to pull a large V-shaped blade below the surface of the ground to cut the sagebrush. A Merrill side delivery rake was used to move the sagebrush into rows where it was burned by hand. It was a dirty, hot and tiring job working in the loose, sandy ground, day after day until the 160-acres were cleared. Sometimes they set an old tire on fire and pulled it behind the tractor setting the brush on fire. It was hard to find old tires for this purpose. One day, Ruby saw a pile of tires along the side of the road near a corner. Thinking someone had dumped them, she proceeded to load them in the pickup and haul them home. It wasn't until later, she discovered operators of the large tractors with metal cleat tracks used those tires to cross the roads so it wouldn't harm the road surface.

Vern Townsend, a sprinkler salesman, told them about a man from Ephrata who loaned money for items such as sprinkler systems; that is where they financed the complete sprinkler package. The method of irrigation was a handline sprinkler. There were eight lines of 4" pipe, one-eighth of a mile long that had to be moved twice a day. A few years later the handlines were replaced with wheel lines, then finally they leveled the farm for furrow irrigation. In the mid 1970's, a circle pivot was installed which was more efficient and saved labor and water.

The second unit wasn't in their name when they cleared and prepared it for irrigation in 1954, but people's word was good then and they were certain the owner would carry out his promise to sell, which he did. It was ready for a crop in 1955. An old wagon road cut across the ground, they never knew where its destination or beginning was. Several crop seasons passed before not being able to see its original path.

Friends and relatives in Idaho told Earl and Ruby that they could not succeed starting without anything, that they would be back.

Ruby said, "After some of the wind storms hit us, I thought maybe they were right." But through all the development, winds and work, neither one of them ever regretted the move, they had found what they were looking for.

Beside the Healys, there was another old time family down the dirt road to the north. Ted and Bessie Nave were dryland farmers and were now adapting to irrigation. It wasn't long before others started to arrive on their land. One of the first was Parley Jenks and family moving to the west of them on ground they acquired in a government drawing. The Hirai family, the Reeds, Ashbys, Wisers, Manleys, Petersons and Riggs, all settled within a mile. Most of the immediate neighbors had prior farming experience and nearly all of them survived the development period and prospered. There were many in the Basin who were not so fortunate.

Haidi and Shegeko Hirai lived next to them and were great neighbors. One time, Haidi bought a new car. Shigeko stopped to visit Ruby and parked in the driveway. A little later, Earl came in from the field on the Ferguson tractor for fuel, the fuel tank was near where Shegeko was parked. To reach the tank, Earl stopped the tractor behind the new auto. Ruby and Shegeko were looking out the window as Earl, leaving the motor running, stood off to the side of the tractor, which he thought was out of gear, and hit the starter button. The Ferguson started all right, but it was in gear and the women helplessly watched the tractor lurch forward into the rear of the brand-new automobile. But Haidi didn't get upset.

Haidi's mother used to visit them occasionally. She only knew a few words in English. She tried to talk to them, but they usually didn't know what she was trying to communicate. The only thing they both understood was, "no savvy." Hiadi would often stop at their house after the days work was done. They would visit for a while and everyone was tired, Hiadi would go to sleep while sitting there when everyone else wanted to go to bed. But Ruby and Betty would get even in different ways; once they passed the field where Hiadi was moving siphon tubes on the ditch bank. His arms were holding a large bundle of tubes. They started honking and waving and they didn't quit until he finally waved back. In the process of trying to wave with one hand, the siphon tubes got out of control and went flying everywhere. They giggled all the way home.

They were filled with excitement and anticipation as the first test water came down the irrigation lateral the fall of 1953. The first crop year was 1954. They planted 10 acres of wheat and about 70 acres of Red Mexican Beans. The beans were well suited for the climate and soil. Ruby said, "When the young beans needed water, they turned a dark color, the leaves came together and pointed upward, as though praying for water." They didn't need much fertilizer and yielded 30 sacks clean per acre. The price was seven dollars a hundred, very good for the time. They thought they were rich.

Beside helping with the field work, Ruby got a job at Jay Harper's warehouse sorting beans for the first few years to help pay expenses. Agriculture commodities expanded rapidly in the area. Harper's warehouse was filled the fall of 1954. A pit was dug behind the warehouse and lined with plastic to hold the extra beans. Earl and Ruby's crop was the first to go in it. They stored very well.

The first building was a well pump house because water was a priority. They built their home in 1956. Harold Beckemier, a carpenter from Quincy, along with his son, Hank, did the construction. A few years later, Harold also helped build the concrete block ends for underground potato storage. It was one of the first storages in the Quincy area and they started to raise Russet Burbank potatoes. The poles used in the storage, were cut in the hills above Leavenworth.

They used the little Ferguson tractor to drag the logs to a clearing where they were loaded on a flatbed truck belonging to Everett Lobe of Quincy. After Lobe delivered the poles to the farm, they were peeled by hand before beginning the construction of the storage. Two of Earl's brothers, from Twin Falls, came to help put up the timbers, stretch mesh wire across the top and put a layer of wheat straw on for insulation before finishing the surface with sheet metal.

One of the first potato buyers was a small processing plant in northern Idaho. Later years brought processors from southern Idaho and the Tri-Cities and finally Brown and Kelly in Quincy which ended up being the present Lamb Weston plant of today.

The first three or four years they planted the Red Mexican Beans. After the first two years the yield's became lower because of root rot. They started to raise alfalfa, wheat and clover seed to replace the beans. The longer the area was farmed, the more disease and pests invaded the crops. When they started, no chemicals were used at all, but over time they became a necessity because without them, there would have been no crop.

Earl tried to get relatives to move to the Basin and farm, but very few wanted to leave Southern Idaho and come to that sandy and windy place, leaving family and friends behind. They did have a lot of company, however, and they all enjoyed the fresh and delicious cantaloupe and watermelon that grew on the ditch banks wherever a seed was thrown. Eventually, several of his nephews did move to Washington, but only three stayed.

It was 1955 when their son, Robert and his wife Agnes, started farming nearby after serving with the U.S. Air Force. Earl and Robert bought a self-propelled Owattona swather. It was only the second one in the area. Earl learned to operate it while unloading it from a railroad flatcar in Pasco where they had it shipped from the factory. He didn't really know what made it go forward and backward. After starting the engine, he had it started in reverse and the rear tire was half over the side of the flatcar before he was able to go forward and onto the dock. The swather was used on hay and wheat straw. They also did custom pea, hay and straw swathing for the neighbors to help pay bills.

Earl was elected to the first National Potato Promotion Board. Their meetings took place in their office in Denver. He was a real advocate and spokesman for the board and loved the friends he made from all over the country while serving with them. He was a natural for the Promotion Board because wherever he went he elaborated on the Columbia Basin being the greatest potato growing area in the world and he had the pictures and facts to prove it.

Both Earl and Ruby became active in the Grant County and State Farm Bureau where much time and effort were put forth to improve local, state and national concerns dealing with agricultural problems. Earl, Orville Child and Hank Thompson led the effort to form the Washington State Potato Bargaining organization.

Earl and Ruby did all the work on the farm. Some of the tasks were plowing, harrowing, corrugating, cultivating, hand weeding, and moving forty foot long sprinkler lines by hand. Later with changes in irrigation methods, they both learned to set siphon tubes to get the water out of the ditch into the corrugate. They dug corrugates with a hoe at the ends of the field where the tractor couldn't get close enough. They combined the beans, raked and baled the hay.

During potato harvest, Earl ran a two-row harvester and Ruby picked vines on the back in the dust and sometimes winds. When not picking vines she drove a truck along beside the harvester as the digger boom delivered the tubers over the side board and into the truck until loaded. She would then drive the truck to the storage, back it down the slope and into the narrow opening of the underground, darkened storage. The boards were pulled one at a time to expose the tubers to a continuously running link chain. The chain carried the tubers out the rear of the truck bed onto the piler that put them into a bulk pile up to thirteen feet deep.

Earl and Ruby, along with several other farmers, Johnny Morris, Paul Hirai, Bob Holloway, Ed, Larry and Dean Pearl and Lorin Grigg, joined Roy Leach, a salesman with export experience, in forming an export company called Royal Foods. The only problem, they were about 20 years too soon. After lots of effort, Roy finally got permission to ship dehydrated potato products to Korea. Washington State processors would not sell any to them. Finally an Idaho firm agreed to make the product to the Korean specs. Just as things were ready to sign, the Korean President was assassinated. Everything was called off. Royal Foods never recovered and went out of business.

They were also part of a group of farmers and agribusiness people who started a research and development group called Quincy Chemical Research. The goal was to use potatoes for something other than to eat. The group originally consisted of 20 people who funded the research, that number grew over the years. The project was started in 1977 and is still active in the year 2002. Some of the entities the group has worked with are Archer Daniels Midland, Battelle Northwest, Industrial Development Corporation of South Africa, Simplot Foods, the Icelandic Government, British Sugar and many others. Earl didn't live to see a plant built from the technology developed, but Ruby may yet see their efforts succeed, with possible plants in not only the U.S., but also in South Africa and Iceland.

The jackrabbits were abundant to say the least. They came into the fields at night by the hundreds. There were jackrabbit drives to cut down on the population. Quincy was the only place I'm certain that held an annual "Jackrabbit Race" at the high school football field. The school named their sports teams "Jackrabbits" and the tradition continued for half a century so far. Every so often, a high population of rabbits would catch a disease and almost die off before making a comeback. That may have been part of the reason for the end of the rabbits, but the development of so much ground in the first 20 years of water took most of their habitats away forever.

Earl and Ruby made a lot of trips back to Idaho, not only to visit, but to pick up seed or supplies which, for a while, were more plentiful in the more established farming area of southern Idaho. They would work all day in Quincy, leave in the evening and drive straight through to Twin Falls. Most of the trips were made together, but one time Earl went to California with a relative or neighbor. While there, he made arrangements for some flowers to be delivered to Ruby. A few days later, she got the bill.

After getting on their feet financially, they not only travelled to Idaho, but to Hawaii and later to Europe twice. Even behind the iron curtain into Russia, Poland, Belgium and some of the other eastern block countries during the cold war with other Potato Board people.

They were in West Berlin during the period Russia blockaded it from all sides and all food and supplies were taken in by an airlift. They saw the infamous Berlin Wall. They had arrived by air from England. But when they left it was by train. When aboard the train, the Russian authorities separated the men from the woman into different cars. The windows were curtained off and passengers were told not to look out. The trains only went out at night. Ruby peeked out through the window all during the trip that night and could see Russian soldiers on bicycles, guarding every road crossing. During the ride, Ruby and Thelma, Earl's sister-in-law, were in a compartment with three women that worked for NATO. At one point, Ruby couldn't find her purse.

In a Russian controlled area that was serious because all of her identification including her passport was in it. They searched the compartment with no luck; they called the conductor who helped in the search, still no purse. She was now in almost a panic when down the aisle of the railway car came another conductor with a lady that had found the purse wherever Ruby had left it. They were bringing it back, much to her relief.

Thelma was with them in West Berlin to visit her son, Dale, who was stationed in the U.S. military. Dale went with them in a friend's van and toured much of Europe for 30 days. One of the points of interest was Hitler's Nest. Some of the farms they visited had the family living in one end of the house and their pigs in the other end. It looked like close quarters, but the pigs for some reason didn't give off an odor.

When they became a little older, Earl and Ruby bought an acre in Quartzite, Arizona. They moved in a mobile home and spent the winter there along with friends from Quincy and many other states. They loved the terrain and the view of the horizon, especially in the early morning and evening. But Quincy was still home.

Earl and Ruby's last trip together was to Israel. After Earl passed away in 1986, Ruby still did some travelling. She went with her daughter, Betty and husband, Ray, to Hong Kong, Singapore and Thailand where she rode an elephant.

Ruby also went with Ila Child to Alaska and Canada. She flew many trips across the western U.S. in her son's single engine Beech Bonanza. The destinations being Twin Falls, Phoenix, Cody, Wyoming, Santa Fe and Bakersfield among others. She enjoyed seeing the country side from a lower altitude. It was a different perspective looking at the places such as the Grand Canyon, Canyon lands, Utah and the mountain areas of the western U.S. where she had travelled in an auto.

They have always believed in higher education. Since they didn't have the opportunity to attend college perhaps they realized the importance of higher education and have helped numerous young people attend school through direct gifts, and through scholarships at Big Bend Community College at Moses Lake.

Earl and Ruby were like the typical pioneers, starting with nothing, worked hard, took risks and met the challenges that progress demands. They were rewarded for their hard work with the satisfaction of realizing their dreams, a farm of their own and providing for their security while helping a great variety of people along the way. One of the tests in life, is how many friends you make along the way. Earl and Ruby earned an A+.

Betty Holloway inspecting the first mail box

Earl and Ruby

First home site
1953

First irrigation water delivery
1953

Earl harvesting first potato crop

Building the first potato storage

Earl and Ruby, 1980 Potatoes

Top row: Howard Cope, Ed McCullough, Lawerence Wilhelm, Claude York, Norman Cope, Earl Holloway.
Bottom row: Myrtle Cope, Ann McCullough, Babe Wilhelm, Viola York, Betty Cope, Ruby Holloway.

Earl and Ruby

Roy Hull

Making a Difference

Roy Hull is a quiet man, but after the silence of a mountain top without seeing or talking to anyone for 45 days, it was boring enough to almost drive him crazy. He wanted interaction with people and being involved in projects that make a difference. He decided a life of solitude was not for him.

Roy's start in life occurred 25 years earlier when he was born on November 26, 1913 at Whitney, Idaho. His father was a county assessor in the south east area of the state. When his father died of pneumonia in 1916, it left his mother at the age of 34 with six children to raise, three girls and three boys, Roy being the youngest. After attending his first year of school in Whitney, the family moved to Logan, Utah. He attended public school at Logan. Upon graduating from high school, he entered Utah State University.

While attending a school dance one evening in 1934, Roy met Merle Anderson, an attractive young lady from Hirum, Utah. Merle was born in 1914. She graduated from high school in Hirum and enrolled at Utah State in 1932. The chance meeting began a two year courtship resulting in a marriage at Blackfoot, Idaho, on August 28, 1936.

Roy dropped out of school for lack of funds for two years. He worked at various farm jobs, including milking cows for a year and picking potatoes at Shelly, Idaho. He saved enough money to reenter Utah State and graduate in 1938 with a B.S. Degree in Forestry. Merle graduated from U.S.U. and taught school for one year before deciding to be a full time homemaker.

Roy's first job was for the Forestry Service. They sent him to B Point, Idaho, a mountain top located north of Lake Pend'Oreille. When his 45 days on the mountain was over in October of 1938, he told his boss, "That's enough for me, I'm not staying here anymore." He left the Forest Service, boarded a bus and headed home to Logan, Utah, and to his wife and their baby daughter, Dianne.

One day Mr. Wrigley from the Agricultural Stabilization and Conservation Service stopped to see Roy's brother Robert. Roy said, "My brother isn't available, will I do?" Mr. Wrigley said, "Sure." He was hired as a fieldman for the Cash County ASCS office. Roy worked there until 1945. Meanwhile, it had been a productive period for the Hull family, by then they had another girl, Judy, and on January 31, 1945, were joined by twin boys, Richard and Steven. As soon as the boys appeared on the scene, Roy thought to himself, "Holy smoke, I've got to get a higher paying job."

Amalgamated Sugar Company had a fieldman opening. Roy applied and got the job. The plant was located 16 miles north of Logan. He covered Cash Valley, Smithfield, Cornish and Lewiston areas. Things were going fine until the yields started to decline because of a lack of crop rotation, some fields had sugar beets for 20 years in a row, allowing the nematode to flourish. After a couple of moves, the company built a house for them and they moved into it in 1947. Roy and the company were very satisfied with each other.

The first Roy and Merle heard of the Columbia Basin was when the company mentioned there might be a job opening in Washington State. In November of 1952, they drove up to the basin to look around. When they got back to Utah, the president of Amalgamated,

Arthur Benning, asked Roy if he would like to move to the Columbia Basin. Roy said, "Sure."

The Hull family arrived in Quincy January 31st of 1952. There were only two new houses available in Quincy. The one they chose must have been a good one. Roy is still there 49 years later. The big influx of new people in the area was just starting, many of the later arrivals weren't so fortunate and couldn't find housing. Roy thought Quincy looked a little 'tacky' when they first arrived. The weather was mild enough in February to pour cement sidewalks and a drive way before spring came.

Roy had two job priorities. The first was to contact the Northern Pacific Railroad (which later became the Burlington Northern). He worked with a railroad man by the name of Richardson to arrange for rail spurs at Quincy and Winchester. The second task was to develop dump yards and the piling facilities at both sites. He was able to accomplish both during the first summer and was ready to receive the sugar beet crop that fall. To comply with all the red tape and permit hassles would take five years to do the same thing in the year 2000.

At the same time he was arranging the sites, he was also contracting beet acreage to local farmers. They planted 1,800 acres the first year. One grower was Don Davidson, south of George. His crop came up fine, but then seemed to deteriorate. They were quite puzzled until they drove out to the field at night, in the headlight of their pickup, the field seemed to be alive and moving. The jackrabbits were thick and feeding on the new plants. There were jackrabbit hunts held in the earlier days trying to reduce the population. Between disease and the sagebrush being torn out for farm development, the rabbits eventually almost disappeared.

Starting in the fall of 1953, lines of trucks at the dump site could be seen waiting their turn to unload. The truck drove onto an elevated platform next to a pit with a conveyer at the bottom which delivered a steady stream of beets to a large piler that stacked the crop in a pile 25 feet deep and 60 feet wide. The truck beds were hinged on the side. After dropping the side board, the bed was lifted from the opposite side allowing the beets to slide out the low side and into the receiving pit. After dumping the load, the truck pulled forward and the excess dirt that was separated from the beets was dropped back in the truck. It only took a couple of minutes to be unloaded and on the way back to the field where the excess dirt was hand shoveled off before the next load.

Roy was well suited for his job. First of all, he genuinely liked working with people. He treats everyone with respect. After dealing with him, the farmers and business community found he was dependable and competent, he could be counted upon to complete a difficult task.

Roy planted 10 acres of beets north of Quincy by the Auburn Pack Feed yards. He used it as a demonstration model to show growers and prospective growers how to grow the crop and what the area was capable of producing. Farmers from out of state appeared at his home inquiring about the possibility of getting contracts if they moved into the basin. They inspected Roy's ten acre field for quality and size of the crop. People came from many states, but mostly Idaho, Utah and Colorado. They wanted to know how the beets would grow and yield. They dug up almost half of the ten acre field to check them out. Some of the growers were Schwints and Greenwalts from Colorado, Chris Hyer, Wayne and Emery Wiser and Bob Hammond from Utah.

Everything was going so well, it was a little boring until the fall of 1955 when the winter settled in early. The morning of November 11 started out very nice, but as the day went on, the temperature got colder and colder. That was the year Bill Petrak and Gordon Fullerton went hunting in Montana. Before leaving, Gordon had dug a few acres, enough to pay back Amalgamated for money they had advanced for growing the crop. (It was common practice at the time to help finance growers for seed, fertilizer and even equipment and harvesting.) When the hunters got home, harvest was over for the year. Two-thirds of the entire beet crop was frozen in and never harvested.

About 1960, U & I Sugar Company took over ownership of the basin plant at Moses Lake. Amalgamated offered Roy another position in their plant in Caldwell, Idaho. He declined the offer and stayed on with U & I until 1963. Roy started farming with Leo Sheehey and Ken Murphy about 1955 and when he left his position with U & I, he farmed full time.

At the peak of his farming operation, he raised 900 acres of crops. He had sugar beets and field corn along with a variety of seed crops such as wheat, radish and parsley. Roy included cattle in the farm operation. He started with cattle when he first arrived in 1953. They fed on crop residues of beets tops and corn stalks, then were finished in the feedlot. All of Roy's ground was watered with gravity flow irrigation until installing a circle system in 1990 and another one in 1998. Most of his farming has been in the Winchester area east of Quincy along the highway to Ephrata.

It was a sad day for Roy when U & I shut down operations at the Moses Lake plant in 1979. The sugar beet industry had been a big part of Roy's life work until that time. It was a great economic loss for the basin area. On the surface, it was as if all the work and effort Roy expended for 23 years had been wasted. But the beet industry had created a solid economic base for many farmers and helped them to survive the transition into alternative crops.

Brown and Kelly had a potato warehouse in Quincy. Percy Kelly also farmed next door to Roy. One day in 1963, Percy approached Roy asking for help to raise money to convert their warehouse into a potato flake plant. They needed $120,000 for processing equipment. Roy said, "Sure, I'll help you." Before he knew it, he was right in the middle of the project. Roy, along with Orville Child and Bob Gilcrist, contacted farmers trying to raise money for stock in the new company. By the time they finished they had more than $200,000 of stock sold. The new Brown and Kelly Flake Plant became a reality in 1963.

Roy served on the Quincy City Council for 12 years. It was during that period the processing plant was being organized. In order for the plant to proceed, it was mandatory to establish a system to handle waste water from the operation. Roy's experience and leadership proved to be vital in getting through the permit system and working with different agencies to get approval for the settling ponds southwest of town. That system has been a necessity for the later growth of processing in Quincy and the economic impact of the Quincy valley. The city paid for the development of the ponds.

The new Brown and Kelly board members included Percy Kelly, Aaron Brown, the plant manager, Orville Child, Bob Gilcrist, John Baird and Roy. After being in business for a while, they found it tough competing with the large operations like Simplot and Ore-Ida. Lamb-Weston approached Percy with an offer to buy them out. A vote was held and the result was to sell. The vast majority of the stockholders signed their stock over and the sale was completed in 1965. The stockholders received 10% profit on their investment. Normally farmer owned enterprises don't turn out that well. Everyone was happy except Brown. He sued Child, Gilcrist, Baird and Roy for $850,000. After two years, a ruling declared they had a majority and the right to sell. Lamb Weston purchased the ponds from the city and took over the operation and maintenance. Later on, Simplot bought into the program with Lamb.

In 1974, the mayor's wife become ill and he resigned. The Quincy city council appointed Roy to fill out the term, but he didn't run for reelection. Between his work on the settling ponds while on the city council and setting up the original receiving areas for the sugar beets, Roy was instrumental in two of the largest economic building blocks in Quincy Valley history after the arrival of water

Roy and Merle were fortunate to travel a great deal. Along with friends Duane Marcuson and his wife, they have been all over the world. Some of the their journeys have been aboard ship through the Panama Canal and down to Costa Rica. They have flown to Japan and New Zealand. While they were in Australia, they attended a play in the Opera House and had a look at the primitive Outback. Other destinations have been London and several times to Hawaii and Mexico. Three weeks were spent traveling in Africa. In the Serengeti Crater of Kenya they viewed huge herds of antelope and other African wildlife including an elephant, lion, zebra, rhino, hippos and giraffe. The lions were so close to their Land Rover, Duane, with camera in hand said, "Roy, get out and stand beside them, people aren't going to believe this." Thank goodness Roy doesn't follow instructions very well.

When they went to Zimbabwe, it was still called Rhodesia. It had been difficult to gain entry because of political unrest. The English were banished from the country. The English have been there so long that English is the official language. The whites eventually reentered the country and developed large farms which the present day government, headed by President Mugabe, is moving the landowners off and giving small acreage to black veterans.

In South Africa, they visited the modern city of Johannesburg. They went on to Capetown at the southern tip of the continent where Table Mountain stands high over the city and can be seen from ships far off the coast. They traveled down the empty white sand ocean beaches from Capetown out to the Point of Good Hope to visit the rocky and windy site of the lighthouse. It was climaxed by taking pictures on the rocky shoreline when Merle, wading in the surf, slipped and the camera went flying through the air to be claimed by the waters where the Atlantic meets the Indian Ocean.

Roy has been active in the LDS church throughout his years in the basin. He has been on the High Council and a teacher of the Gospel Doctrines. He said, "Our religious belief has been a big part of our life." He's also been a member of Farm Bureau for many years.

Merle passed away July 9, 1989. She and Roy had six children. Marianne and Richard operate a pillow shop in Quincy. Steven is teaching at Brigham Young University. Judy lives in Orem, Utah. Dianne is married to an ophthalmologist and living in California. Ron operates the family farm at Quincy.

Roy is happy they came to the Columbia Basin, he said, "The Basin has been good to me." That works both ways, Roy has been good for the Basin. He came down from the mountain top to live in the valley. Through the years, the quiet man's actions have been a lot louder than words.

Roy and Merle

Roy Hull

Dean and Catherine Moore

Join The Army and See The World

The bullet entered his back just missing his shoulder blade. He felt his whole side go numb slowly. Raising the hydraulic controlled disc, he headed the tractor across the field for the house at full throttle. He could feel the warm blood seeping downward inside his shirt. This can't be real, he told himself. After completing 35 missions over Europe during the war without being harmed, Dean had just been shot on his farm in peaceful southwest Idaho.

They turned their ground into a good farm with lots of work and good management. Dean worked at John Deere Agency during the day and did his field work at night. Not all of the neighbors were that ambitious, in fact one nearby family was more than a little jealous of Dean and his wife Catherine's success. They claimed Dean had moved a fence over on their ground once, but they settled it without any real problem. One day the neighbor boy came by to inflate a low tire on his tractor with Dean's small air compressor which Dean hooked up for him. Later that same night, Dean was discing a field with his John Deere tractor in preparation for winter wheat. As he made his turn at the end of the field and started back, he felt something enter his back and out the front of his shoulder.

He was losing a lot of blood when he shut down the tractor at the house. Carolyn, his daughter, was waiting for him to come in from work and was upset when she heard her daddy tell mommy he had been shot. She was only five years old, but knew something was wrong. Dean sat down on the porch and leaned against the wall. Catherine told him to get in the car before he got too weak to manage by him self because she wouldn't be able to lift him if he passed out. He got in the back seat and his wife drove to a neighboring house for help. She was so shook up she didn't trust her own driving on the way to the doctor in Homedale, the nearest town. The doctor took him to the Caldwell hospital for treatment and X-rays. They couldn't find a bullet in him. Catherine picked up his jacket and found two holes, front and back, the bullet had passed on through. He received a shot of penicillin every four hours for three days. The bullet didn't hit a bone, but he couldn't work for a month.

When the Sheriff went out to check, he found the front and back windows of the house had been shot out after the family left for medical help. Where Dean had been sitting on the porch, they found two bullet holes in the wall. The caliber of the bullet was a .22 long-rifle. Catherine told the authorities her guess of the identity of the shooter. Her guess turned out to be correct and they charged the seventeen-year old neighbor boy. The boy's brother-in-law was an attorney. He got the court to place him in a mental institution, but he was out within a year. However, the family never bothered them again except to ask if Dean was going to sue them, which he didn't.

Dean C. Moore was born near Idaho Falls, Idaho in the year of 1919. His father and grandfather were farmers and it was natural that he would follow in their footsteps. The Idaho Falls and Ririe, Idaho area had diversified irrigated farming. Potatoes, wheat, feed grain, alfalfa and livestock were the basic crops. It was enough to keep a young farm boy occupied. He had chores such as irrigating, digging out corrugates, weeding crops, cleaning out corrals and a hundred other jobs necessary during the farming season.

His first years of school were at a two-room school house at Stanton, six miles south of Idaho Falls. They later moved to Roberts,

Idaho for a few years where he attended a one grade per room school. Ririe was a dry land farming area. It was during the depression years of 1933-34. They not only had the depression to worry about, but the rainfall was scarce and the wheat price was twenty cents a bushel. Pigs were only two or three cents a pound. Things were not working out financially. In 1937, Dean's dad rented 160 irrigatable acres three miles east of Idaho Falls. It was an old sheep farm headquarters. After a few years, they were doing better, the depression was ending and they moved back to Roberts. This time they bought a very good 160 acres of loamy volcanic ash soil under irrigation.

They pumped the water out of a canal for watering the crops. It was the first experience with electric pumps. The pump lifted the water out of the canal and into gravity flow ditches. Electrical power was becoming available in the rural areas and the Moore family had it installed in three different houses they lived in during the 1940's. They bought their first refrigerator in 1939. Most of the houses had no indoor plumbing. They carried water in from an outside well.

Dean went to Roberts High School for three years and his senior year was at Idaho Falls where he graduated. He spent one year at the University of Idaho in Pocatello until he ran out of money and quit. He worked hauling coal from western Wyoming in the winter with a ton and a half Ford truck they owned. The route he covered was 40 miles of pavement, 40 miles of gravel and 40 miles of forest service road each way. He delivered coal to friends and neighbors. U and I owned the Lincoln Sugar factory at Idaho Falls where they hauled beet pulp, loading and unloading it by hand. The haul was from the factory to Monte View, forty miles northwest.

The year of 1941, Dean had the opportunity to ride to California with a neighbor and stayed with an uncle in Los Angeles. He tried to enroll at University of California but $400 for out of state tuition was more than he could afford. The next stop was Los Angeles City College. For $5, he got a student body ticket, a pair of swimming tights and towels furnished. He enrolled for the quarter and returned home to Idaho for the summer work on the farm. He registered for the draft October 16, on his twenty first birthday. Pearl Harbor was attacked on December 7th, he joined the Army Air Force in January of 1942.

Dean joined the service as a mechanic. He was sworn in at Fort Douglas, Utah and shipped to Shepard Field, Texas. From Texas he went to Chunute Field in Illinois for aircraft mechanics. The class was almost completed when they shipped him to Buffalo, New York, to study P-40's. The military was phasing the P-40 out, however some were still in use by General Chenault. They sent 200 mechanic students, including Dean, onto another P-40 training field in Florida. They were phasing the P-40 mechanical training out there too. The men weren't doing much of anything until some twin engine, P-70 night fighters arrived. Training on the P-70's was a hot job. They were painted black causing work conditions in the plane to be hotter than normal. It was hot enough to soften a new type of latex fuel line. They installed the lines because they were bullet proof and self sealing if hit by bullets or shrapnel. When the plane headed down the runway at full power and put high pressure on the lines, they had fuel leaks all over the place.

Dean decided to try for an aerial gunner position when he was told the qualifications were the same as for a pilot, why not try for that? He passed the test. He was among the first group who all had college educations. They shipped him first to Miami, than to Mobile, Alabama, for more advanced schooling in mathematics among and other other piloting courses. It wasn't all studying. A second lieutenant had them running up and down a steep ravine near the school each day. By the time they were ready to move on, they were in great physical condition.

From Alabama, they went to Lackland Air Base in San Antonio studying code, physics, and more mathematics, but no navigation. The next stop was Muskogee, Oklahoma, to train in a Brazilian built PT-19. The PT-19 was a low wing, fixed gear, two place, single engine plane that could fly inverted. Civilian instructors taught them, four per group. Once just before soloing, Dean had just finished his flight and the next trainee took off with the instructor. The new pilot was supposed to be flying level, but was nervous and not performing well. The instructor, to get his attention, pushed his duel control stick forward suddenly. That caused the plane to dive. When the instructor looked in his rear view mirror, he saw the student raising up out of the cockpit. The trainee had been so nervous, he hadn't put on his seat belt, worse than that, he had forgotten his parachute. Fortunate for the trainee, he was able to stay aboard when the instructor levelled off. Back on the ground, the whole group received a well-deserved lecture.

The next stop was Coffeeville, Kansas where they flew the BT-13. Dean said he was doing well on his flights before taking a week of leave and another week of inactivity because of weather. After the layoff, he said he couldn't do any thing right. They teamed up to do instrument practice. On one flight, Dean was the observer while the other pilot flew under the hood flying instruments. Dean put the plane at different positions and his companion was to regain control from the unusual attitudes. The other pilot had been doing well when Dean put it in a steep bank and compressed the high rudder. The plane suddenly flipped, throwing all the gyro instruments off. They were at seven thousand feet when they started. By the time Dean got the other pilot to release the controls so Dean could visually straighten the aircraft, the elevation was 5000. Dean didn't try that particular maneuver again.

Pampas, Texas, was where Dean got used to the twin engine PT-17 made by Cessna. He was doing well and enjoyed the flying, including the instrument practice. They were going to assign him to fly a B-26 and shipped him to Dodge City, Kansas. Before he even got off the plane at Dodge City, they reassigned him as a co-pilot of a B-24 and off they went to Lincoln, Nebraska. In Nebraska, crews were put together and sent to Pueblo, Colorado, for B-24 flight training. At the conclusion of training Dean's crew went to Topeka, Kansas, to pick up a new B-24 and flew to Newfoundland and Goose Bay.

They called the B-24 the "Liberator," Dean called them, boxcars with motors and wings. They had 4800 horse power in the engines. En route the generator went out and number 4 engine quit. They could get along with three and continued on to Goose Bay for repairs. There was 20 acres of cement at Goose Bay covered with B-24 and B-17 bombers. Everyone was busy, meanwhile, they waited, trying to get someone to change the bad generator. The night before they were due to depart the next morning at two o'clock, Dean went to the supply sergeant and asked if they had a generator and a special wrench it took to put it in. The sergeant said they had both if he would sign for them. Armed with the new generator, special wrench, a pair of pliers, some wire and Dean's earlier mechanical training, he went to work. Two and a half hours later, they had it installed and they left on time. It had been a long night, but they set the autopilot and the navigator did the rest while Dean slept a third of the way to the Azores.

The Azores runway was rough and rolling. The B-24's nose wheel was sprung on landing. It was no problem on takeoff, but when they were in landing pattern at the next refuelling stop, the gear wouldn't go down. Dean and another crew member went into the lower hold where the gear came up into the body of the aircraft through a 3x6 foot opening. They braced themselves and pushed down on the landing gear with their feet with enough pressure that the gear finally lowered and locked into position for landing. The crew had the Italians work on the gear. The first thing the Italian mechanics did was drop the plane on its nose, it was six months before the practically new aircraft would fly again.

Dean's crew left the plane and rode a truck onto Venosa, Italy, to the group headquarters of the 485th. They were assigned to the 830th Squadron and given a bad time for not arriving with a new plane because bombers were badly needed. The crew arrived the last of August 1944 and were flying a replacement B-24 by the first week of September. Many of their missions were over the Alps with a destination of Vienna, Austria. Other targets were in Poland, Hungary, Germany, Italy, Czechoslovakia and Yugoslavia.

On one of the early flights, another crew, flying off their right wing, lost an engine. They all continued on and hit their target. The crew of the crippled B-24, knowing they couldn't make it back, headed for Yugoslavia. The chances were 50-50 that Tito's men would find them before the Germans did. The plane landed in the rural hill country of Yugoslavia and Tito's forces immediately recovered the crew. The same evening, an allies' DC-3 landed with supplies for Tito. They took the crew aboard and they were back on base by night fall. That particular pilot refused to fly again, but the rest of the crew did.

One pilot was Jewish and if he were ever shot down and captured by the Germans, his fate was sealed. They graduated Dean from co-pilot to pilot and assigned him to take the Jewish pilot's place. That plane had a history of running short of fuel on the long missions. It wasn't long before Dean found it wasn't the plane having problems, it was the navigator who always wanted to turn back for various reasons. Dean informed him that on his plane, they were going over the target. From then on, all missions were completed.

The B-24s left contrails across the sky. So many planes were flying at times the contrails caused visibility problems, not unlike clouds that could extend downward almost to the surface. When that happened, the group couldn't see to fly in formation and would break up and scatter. One of those days, Dean circled down to 3,000 feet above the ground before breaking out in the clear. The elevation was lower than the mountains on the coast they had to cross on the return trip. They were looking for a certain river to follow that threaded its way through to the coast. They spotted the flight's lead aircraft. It had found the river and Dean just followed along to home base.

During one trip over the Alps at 17,000 feet, they were heading for Vienna when his number three engine gave out a loud noise and started shaking violently. Dean feathered the engine quickly and shut the fuel off. If they turned back, they would be alone and the Germans would be certain to send a couple of fighters after them. They elected to go over the target and were able to keep up with the formation. However, with one prop dragging, they became short on fuel. A British airfield was on the return route near the Adriatic Sea. When contacted by radio, the British didn't want them to land, but being low on fuel, they had no choice. Reluctantly the British gave them permission to set down. They received red carpet treatment.

When they got out of the plane, a squad of soldiers with fixed bayonets met them. The British weren't taking any chances.

The crew wasn't allowed to take off again with only three engines. Dean's earlier mechanical training again came in to play. He inspected the bad engine and found one spark plug laying in the bottom of the cowling along with a crescent wrench. Three other spark plugs were loose and almost out. Some mechanic must have been interrupted and forgot to finish tightening them. After tightening the plugs, the engine ran great. Nevertheless, it was too late to leave and they hadn't received enough gasoline from the British yet to make it back. Their hosts gave them luxury accommodations for the night. It was a bombed out building with no roof and an armed guard to watch their every move. The meal the British furnished that night was a cup of tea and a crumpet. The next morning, they were started out right with another cup of tea and another crumpet. At least they gave them enough fuel to make it home. Engine number one wouldn't start. Number two started and with it still running, Dean took the solenoid from number 2 and put on number one. That did the trick, it activated the priming and they were on their way. By the time they arrived, they had been declared missing in action.

Dean was fortunate to complete 35 missions without being shot down. The B-24s were grounded for repairs from enemy fire quite easily. They delivered the payloads from altitudes up to 28,000 feet to reduce damage by anti-aircraft fire. At a height of more than five miles, the artillery shells took almost a half minute to reach them. By flying an erratic zigzag course, the enemy gunners had to take a wild guess where they would be when the shell exploded. The evasive action and having friendly fighter planes flying escort, kept losses down. The famed Tuskagee Airman escorted them part of the time. They were a squadron of U.S. fighters piloted by all black pilots who compiled an exceptional record during the war.

After the war in Europe, they sent Dean to Santa Anna, California, where he was given the opportunity for discharge in May of 1945. Meanwhile, his father had moved to Homedale in southwestern Idaho. His dad needed help on the farm and Dean joined him and rented another 40 acres for himself and put in potatoes. The crop made enough to buy a tractor and they bought another 160 acres. The place was sandy and rundown, but they built it up through good cropping and management practices.

Catherine Curtis was born in Given Springs, Idaho, a small settlement on the banks of the Snake River where it separates Idaho and Oregon. Her parents moved to Melba to raise sheep. She graduated from Melba High School. Raised as a country girl, she loved dogs and horses, including wild horses they caught roaming the area of western Idaho. After high school she attended a business school in Colorado Springs, Colorado. After graduating, she returned to Idaho and went to work for King Meat Packing Company in Nampa. On a visit to see her brother in Homedale, she attended a local dance. There she met a brown-haired, handsome young man with greenish-blue eyes. He was a local farmer who had been an aviator and recently returned from the war. Before the dance ended, the young aviator was smitten by the attractive young brunette from Nampa.

Dean and Catherine were married in 1946. Four years later, the same year their daughter, Carolyn was born, they sold the farm and purchased another sandy acreage that had been partially farmed. They rebuilt the existing house and levelled the land. When they finished working the place over, it was a very good farm. While developing the new place, they heard about the water coming into the Columbia Basin in Washington State. It sounded intriguing and in 1954 he made a trip to look at 80 acres in Eltopia connected with the GI drawings. The home place had to sell before they could make a move, so they turned the Eltopia ground down. The following year he returned to look at another parcel near Mesa, but they still weren't ready for the move.

In 1952, Catherine was carrying her young daughter in the field one evening on the Idaho farm. While waiting for Dean, she experienced something she would never forget. There was no wind and she could hear two airplanes off in the distance. Toward the other side of the field from her she spotted an object moving slowly through the air. The altitude was very low and it made no sound as it passed within 1,000 feet from where she stood. She could still hear the airplane from the other direction as the strange craft drifted across the edge of the field. It was like two saucers cupped together and had windows around the upper outside edge. It was at least 200 feet in diameter. It was just dark enough to see that the windows were lighted from the inside. The whole scene felt unreal, but she knew she was not imagining what she saw. Dean had his head down cleaning out corrugates in the same field, but with no sound, it didn't attract his attention. After that sighting she always had a special interest upon hearing of any UFO reports.

In 1958, the Federal Aviation Agency wanted more Air Traffic Controllers. They gave pilots preference, especially if they were veterans. He applied and was accepted. They sold the farm. Dean received training in Oklahoma City before moving to Spokane in 1959. It wasn't the Columbia Basin, but they weren't far away. He worked at Fairchild Airbase west of Spokane. They didn't like living in the city and kept watching for a place in the county. Returning from Wenatchee one day, they came up Trinidad Hill from the Columbia River and looked out over the Quincy Valley. They decided right then, that was the area they had been looking for. A year later in 1960, they left Spokane and came to the Quincy Valley.

They bought a farm unit in block 72 near the Winchester Wasteway where it crosses the highway from Ephrata to George. It was seeded to alfalfa and watered by handline sprinklers. The 80-acre unit next to it had a small house converted from a chicken house. Dean rented it from the owner, Clyde Allan, with an option to buy. Another unit to the south was for sale. Dean and Catherine put enough money together to buy it on a contract. The three parcels gave them most of a small valley and consisted of 310 irrigated acres. Most of the lower ground had a natural slope to it that allowed gravity flow irrigation.

Besides alfalfa, they raised wheat, field corn and dry beans. Baled hay was only bringing $16 to $18 a ton in 1962 and the buyers usually threw out a good percentage of it, they only took what was perfect. Cattle were kept to feed the left overs from the hay buyers. He traded hay for 120 head of sheep. They used baled straw for shelter by making walls of the bales and covering the top with plastic. They sold wool and lambs. Dean was surprised, the sheep paid better than cattle.

The winter of 1968 and 69 was hard on livestock. One morning it was 16 below zero and the wind was blowing. The snow in the road was four feet deep. Dean's cattle huddled together in a tight group for each other's body heat. He fixed a discarded hot water tank with heat bulbs under it so it wouldn't freeze. They stocked enough corn silage in the fall for winter feed. If not for the warm water, many of them would have died from the cold. Catherine's father was staying with them. He tried to help with the stock. Nevertheless, he was 77 years old at the time and the temperature was too severe. Dean had to help him back to the shelter of the house. They were fortunate, only three lambs were lost during the cold spell.

Moores were close friends with neighbors, Everett and Dee Thornton. The Thorntons always had cattle too. One day three head of Everett's got loose and wandered in with Dean's cattle. Dean was rounding up his cattle to sell at the Winchester auction when he noticed the three extra head. They were a different breed than his and not knowing where they come from, he loaded them with his own and took them to the saleyards. He instructed the sales people the three were not his and to put them separately when they unloaded. A short time later the Brand Inspector called Dean at home and charged him with rustling. Evidently the stockyard help weren't paying attention when Dean told them they weren't his. Checking around, he discovered they were Thornton's after all and everyone had a good laugh.

After years of irrigation, the underground water table in many areas of the Quincy valley kept raising. That resulted in salts forming on the surface, ruining the ground for farming. To prevent the destruction of land, the Bureau of Reclamation designed and installed underground drainage systems of perforated pipe. Two of those drainage lines crossed Moore's farm and drained water into the Winchester Wasteway. At first, each 8-inch pipe-line discharged a full volume of water, as time went on the amount decreased, but they continue to flow all year around. The drainage systems reclaimed many acres in the Quincy Valley, including Moores.

The Moores quit farming in 1986 and rented the land. A mortgage was still owing to FHA, but the rent covered all the payments and things were going along fine. When FHA found Dean was not farming it himself any longer, they required him to sell most of the farm and pay off the note. They moved near Hermiston, Oregon, bought property and put up a few rental homes. Through the years, Catherine's health gradually failed and she passed away in 1998.

Dean has never been sorry he and Catherine came to the Basin, it was the best move they ever made. With a twinkle in his eye, he said, "Of course, we moved seven times and never left our farm." They had lived in seven houses and mobile homes on the farm during the 26 years. In addition, they improved the ground, added to existing buildings, drilled new wells, built animal shelters and work shops.

Dean was a small town country boy that travelled to more places in the United States and around the world than ten average men. Nevertheless, the dream of being able to own land and prosper in the Columbia Basin was desirable over all the exotic far away places where he had been. He also proved that farming is almost as hazardous as in the middle of a shooting war.

B-24 Squadron over Europe

Bombs were dropped from 26,000 feet.

Airman Dean

Catherine and Dean Moore on their wedding day. Sept 16, 1946

The Brazillian built PT-19

The B-13-A

Catherine on her Western Idaho mustang

John Morris

An Ordinary Likeable Guy

I remember John as an ordinary man who was anything but ordinary. He was one of those special people that everyone likes without even knowing why.

I first met John in a real estate office when we were both young and trying to supplement our farm income in the early 60s. He had a roly-poly build with good natured, moon shaped face. He stood about five foot, eight inches tall. You could feel as well as hear the chuckle in his voice as he would tell about something that to someone else would be a disaster, but to him, just a stepping stone in life. He never cared how you were dressed or if your hair was combed, he looked deeper than that and saw your interests and concerns.

John never complained about anything, weather, crop prices, personal problems, not even the government. If something went wrong, he didn't blame someone else, instead, he used mistakes as learning tools. The real estate sales didn't help finances much so we both dropped it.

He soon became successful in farming with hands on management. He would kneel down in the corn or potato field and analyze the moisture in his hands to see if the crop needed water. He would utilize crop fieldmen for recommendations on chemicals and fertilizer, and made certain that tillage, planting, cultivation and harvesting were done at the time needed. John enjoyed his work and the every day challenges that confront almost any small business.

It wasn't just farming that he enjoyed, he telephoned several friends each week to chat about a variety of subjects. There wasn't a nearby high school basketball or football game that he didn't know what the outcome was. The same thing for the Washington universities and the professional teams.

He loved to play on the commodity markets. One day he would call and say he bought two contracts of wheat and the next week he could call back, laughing at himself and say, "I'm behind about a $1,000.00." For the people fortunate enough to be well acquainted with him, it was a joy to talk and be around his cheerful and positive outlook on every thing. You just felt good after visiting with him.

John was a risk taker financially besides the commodities, he got involved with many projects such as a trading company hoping to create a market for crop sales to the Pacific Rim countries. That company was about 20 years before its time and finally died a quiet death. He invested into a research and development group, made up of local farmers and agri-business. They were trying to make potatoes into something besides food. That was in 1977 and today that effort is starting to bear fruit. Within two years it may create 5,000 jobs in a country where unemployment is at 50%.

During this period he and his wife raised three sons. They taught them to farm and to be good, hard working, honest citizens who are a credit to their community. Also during that time, because of diabetes, he was losing his eye sight while still at the real estate office. He was completely blind soon after, but he kept right on farming.

Over the last several years of his life, the disease took its toll with amputations starting with the toes of one foot, then the other, after that it was the feet, the lower legs and finally, both complete legs. Not once during the whole time did I ever hear a complaint. Nothing but a smile and the latest score from the last Ephrata ball game or how much he had won or lost in wheat futures.

John passed away almost 20 years ago, but his courage and outlook on life makes him very much alive in my memory.

Ken and Dorothy Murphy

If You Don't Like It, Change It

The first duty early in the morning on the dairy farm was to put the cows in the barn. The next for the brash young man was to wake up his boss to help with the milking. His boss didn't like getting up in the mornings and liked to go back to sleep. By the time the young man had brought in the first bucket of milk, his boss would normally be up to help finish the chores. The farmer was a little lazy and as time went on, stayed in bed longer each day. One morning after calling his boss and getting an answer, he went back to the barn. A little later, when he brought two buckets of milk into the separator, the boss still wasn't up. Very impatient with the lack of response, he went to the cistern and pumped a dipper of cold water. He walked back into the house, into the boss's bedroom. He was in the motion of splashing the water over his boss still laying in bed. The man was awake enough to see what was coming and threw up his arm deflecting the dipper splattering the cold water on his wife instead.

That impatient young man's name was Ken Murphy. He was born nearly 20 years earlier at Bantry, North Dakota in 1914. His parents were farmers. He attended a small rural school where each teacher taught four lower grades and the principal taught high school. When a school was built closer to home, there were as many as 20 and as few as five students through the eighth grade and all were taught in a single room. He finished the first eight grades there. His mother had to live away 150 miles from home for him to attend high school at Grafton. Their stay in Grafton was short, he was there only one winter. When March of 1930 arrived, he returned to work on the farm. Including grade and high school, he was in school a total of 36 months. He wanted to go to an agricultural school, but in the middle of the depression year they couldn't afford the cost.

Ken along with a friend, took a farm job near the Canadian border harvesting beets in the fall of 1935. A horse drawn implement lifted the beets out of the ground. The boys took two rows apiece, they grabbed the tops and threw them ahead into small piles. Afterward they threw four piles into one, cutting the tops off at the same time. In the evening, a truck went through the field and the beets were picked and loaded by hand. The loads were delivered to the railroad where they were conveyed into railcars headed for the factory. Each day they worked from dawn to dark and made $8.25 a day, the highest wages they had ever earned. The amount of pay depended on the volume of beets harvested and they made the most of it. They had breakfast before daylight and were at work at dawn. At 11:30, one of them prepared lunch. They both ate at noon and then returned to the field where they worked until dark. At the end of the three week harvest period, both were in good physical shape and had money in their pocket.

Ken's friend had a sister in Oregon and he decided to join her and find work. Ken went home to help his dad on the farm during the winter. The next spring he built pasture fence. In late June of 1936, he boarded a train and headed west to the Oregon coast. He had his friend and a job waiting for him, helping to build an auto-court south of Newport at Waldport. They built seven residences. The complex was located at the bottom of a hill. The water supply was a half mile above the cabins. A water line was constructed of trees that were hollowed out with a four inch hole through them and laid end to end down to the auto-court.

Ken's boss at the auto-court wasn't the brightest person around. The cabins were located along the highway. In 1936, motor traffic was sparse, perhaps a car every 30 minutes. He had a phobia of being hit by one of those motorized vehicles. Sometimes he waited 20 minutes to cross the road.

The area was still a little primitive, the nearby river was still crossed by ferry boat. The bridge was built in 1936, the year Ken arrived in Oregon. Ken worked there a month before moving north to stay with relatives near Salem. He worked at various odd jobs including a dairy and a vegetable cannery. At the cannery there were a lot of men from the dust bowl area of the country. During lunch on a sunny day, he noticed the men from the east sat in the shade, the ones from the coast where it rained so much, sat in the sun because they didn't get the chance very often. He and a friend soon went north again and ended up in the town of Bay Center near Aberdeen.

There was a young lady whose father was part owner of Heath and Cearn's Hardware store in South Bend. She was born in South Bend in 1921. Dorothy Cearns had two brothers and four sisters. She graduated from South Bend High School in 1937 and worked in the local bank for three years as assistant bookkeeper. She attended the Catholic Church in town. Her family was active in the church choir. Ken worked at the Chevrolet garage in a small town four miles away as a body and fender man. His boss lived next door to Dorothy. Ken and Dorothy attended the same Catholic Church and being a healthy young man, became aware of the hazel-eyed brunette that sang in the church choir.

Late in November of 1941, Dorothy left for Marymont in Tacoma to be initiated in the Dominican Nuns. The first 18 months were spent in study and preparation. The last two years were spent teaching at the 3rd and 5th grade school level. A boys' military school was attached to Marymont with grades from 5th to 8th grade. It was a life of service, washing, cleaning and cooking. She loved the children, but found the convent life just too confining. She was there for four and a half years before telling her superiors that teaching was not her calling in life. After leaving Marymont, Dorothy stayed with her sister near Seattle for several months caring for her three young nephews while her sister was expecting a fourth child.

Two weeks after Dorothy left for Tacoma and Marymont, the Sunday afternoon of December 7, 1941, Ken heard the news about the attack on Pearl Harbor. The following morning he told his boss he was going to enlist in the armed services. When he reported in the next day at Aberdeen, the sergeant told him to take eight other recruits on the train with him to Seattle. They were to report in at the Smith Tower building. There were thousands of enlistees waiting in line. A sergeant only had two people helping him sign all the men in. When the sergeant asked the group if anyone wanted to help them, Ken volunteered. He worked for three days before they had time to swear him into the military. Since he wasn't officially sworn in, he didn't get paid for the three days. He probably wasn't so quick to volunteer the next time.

From Fort Lewis, Washington, they were shipped non-stop by train to St. Louis, Missouri. The train was occupied exclusively by G.I. Joes. After arrival at St Louis, they were put in old tents that had been used by the National Guard. The tent tops were very low and had wood floors with half inch cracks between the boards. The material the tents were made of was old and dried out. The sidewalls lacked a foot of reaching the floor. Temperatures went down to six below zero during their stay. In the middle of every tent was a heating stove. It was kept red hot all the time. At night, the guards used any kind of an excuse to stop in by the heater. Three or four tents caught fire and burned in three days before they were moved into barracks. It was only three or four days in the barracks before they were shipped on to training camp.

Ken was sent to Chicago to train as an airplane mechanic. There were 807 students in his class. After graduation, they took 625 with the highest grades and told them, "You are now instructors." They could pick the category of their choice, Ken chose electrical. None of them really wanted to be instructors, a few said, "We aren't going to be any damn instructor." The majority of the men met together and decided, if that is what the Army thought would do the most good, they would do their part and do their best.

Ken was stationed at various camps, mostly in the eastern part of the country conducting classes in aircraft electrical systems. He saw his first jet plane at a base in North Carolina. It was an experimental propjet from Bell Aircraft. He said he had never seen any aircraft climb at such a steep angle before. On the 3rd of May 1942, he left Lincoln, Nebraska, by train for Goldsborough, Carolina, Long Island and Massachusetts for 10 days at each destination for more training in preparation for overseas duty.

His unit was sent to Great Britain and attached to the 9th Air Force. After five weeks of waiting they where sent on the Queen Mary to Scotland. From Scotland they traveled to London by train. After a few days in London they were transported to southern England. The outfit was there for a year before the Air Force wanted them in France. They arrived in Paris five days after the Allies liberated the city. There were a few German snipers left to be cleaned up. Within one day and night they received orders to return to London enroute to the Pacific.

They were taken across the channel to Liverpool and loaded on trucks bound for London. It was night as they headed for the city. They hadn't had a meal for some time and no cafes were open along the way and everyone was hungry. There was a small town where the truck driver stopped to see if they could find any food. When the truck stopped and the men unloaded, they heard a yell. From all over the town women started to appear. There were 50 women surrounding them, there wasn't a man in the whole town. Some of the men and the truck driver didn't want to leave. They didn't get anything to eat and they left as quickly as possible, despite the objections.

They were in London a few days before moving on to the west coast to catch a ship back to the States and then to Colorado instead of the Pacific. Ken was still in Colorado when the war with Japan ended. He was mustered out in 1946 and left for the northwest and his old job at the body and fender shop.

It didn't take long before Ken was calling on Dorothy. They had known each other for a long time. She thought Ken was rather shy, but noticed he seemed to achieve whatever he tried to accomplish. Her side of the story is, while sitting in the choir loft at church during services, she could see Ken in the audience below and thought his pompadour hair style looked terrible and she finally had to marry him to fix it right. His version is he was simply irresistible. Whatever the reason, it was followed by a wedding on August 30, 1947.

In 1948, they moved to Salem, Oregon for three years. Before they left Washington, a friend told them about the Columbia Basin Project, but they soon forgot about it. The news reached them again in Oregon about water coming to the Basin and they decided to check it out. The first place Ken found a job in Washington was at Sunnyside. They weren't there long before moving to Moses Lake in February of 1951 to work on the Goodrich brothers farm.

While working for the Goodrich brothers, Ken and Dorothy were buying some farm equipment of their own as funds were available. For a tractor, they bought an old John Deere with iron lugs at an auction sale. Later they traded it to Mode Snead at the John Deere store in Moses Lake. Mode cleaned it up and used it in the showroom for display.

A few years later in 1954, they moved to Quincy and rented ground. Howard Bargreen, the Ways and Means Chairman of the Washington State Legislature owned some adjoining ground, and they rented his too. Ken tried to get a beet contract from Roy Hull, the Amalgamated Sugar Company fieldman. Roy didn't feel Ken's soil was suitable for beets, but could get some for the following year if he got better ground. The next year Ken got some good ground from Irl Morrison north of town and they received a sugar beet acreage in 1955. Ken, Roy Hull and Leo Sheehey ended up in a farming partnership arrangement for several years.

The Murphys, along with Roy Hull, and Leo Sheehey, bought 15 farm units in a large block of land in the late 1950s on the Royal Slope. They split it up into individual ownerships. There was some problem getting water from the bureau the first year. They had some private sessions with Phil Nalder, the Director of the Bureau of Reclamation, in Ephrata. At first he said they couldn't receive water, but after some negotiations, it all worked out. All the ground had to have sagebrush cleared and part of the ground leveled.

They farmed at Quincy and the Royal Slope until Ken and Dorothy moved to the Royal Slope in December of 1960. The crops on the slope have been beans, corn, wheat, sugar beets, potatoes and for a while they tried raising Christmas trees. Their banker for many years was Harley Ottmar who worked for the Production Credit Association. The PCA made some very large loans that went sour because of some lenient loans without proper collateral. Hundreds of thousands of dollars were lost and not repaid. Regional PCA officials overreacted in response to the losses and the farmers still operating had their credit tighten up. Many good operators found it difficult to get loans, and if they did, they were required to mortgage more than necessary to acquire them. The Murphys were some of many who changed financial institutions.

Ken became well known in Washington State for his participation in the American Farm Bureau Federation. He served for years on the Washington State Farm Bureau Board of Directors as a representative from Grant County. He was Grant County President for two years and on the county board for several terms. Countless trips to Olympia were made to testify at hearings and to visit legislators on issues concerning agricultural interests. Most farmers were too busy with their own work to fight for or against legislation that would affect farming. Ken sacrificed his own interests for the good of agriculture and to battle for what he thought was right.

Ken made friends from all over the state in the Farm Bureau. He was not only active, but quite vocal in the positions he took on issues at the annual conventions. He said there was one lady from the west side of the state who opposed nearly everything that he supported. At one convention after adjournment for the day where they had taken opposite sides, the lady's husband approached Ken and said, "Please don't stop arguing with her, because if you do, she'll be after me when we get home."

Ken and Dorothy have 13 children, 10 boys and three girls. They had planned on an even number of both, but when the count got to nine boys and one girl, they decided it would take too long to catch up. They adopted two young sisters, to help even the odds. Afterward, they had another boy and decided to forget it. Rebecca and Teresa live in Yakima, Bob and John are farming, Maureen, her husband and daughter, live on the farm, Mike and Jim are carpenters, Andy a parachute jumper, Mark an engineer for Battelle Northwest, Larry is in charge of maintaining equipment that loads containers at the port of Seattle, Jerry is an accountant in Seattle and Bill and Kevin do landscaping in Seattle. They also have 26 grandchildren.

Ken is recovering from a stroke that occurred in 1997. He developed from a brash young man who worked hard to accomplish his goals, to an elder statesman fighting for his community and his chosen industry of agriculture. From the beginning, his actions were direct, honest and on the most part, effective.

Harley and Juanita Ottmar

He Lived To Farm

Harley was only 16 when he almost made the biggest mistake of his life.

The U.S. was going full tilt in the war effort against Japan and Germany. Harley's uncle got him a job with a construction crew. They were expanding the runways on the Ephrata airfield while Air Force bombers were making practice runs just over their heads. The wages were $2.45 an hour and that was big money in 1943.

The money was so good he decided to quit school and stay with the job. After a week out of school, he thought to himself, "You know, I had better finish my schooling because I may need that education later in life". With some reluctance, he dropped the job and reentered school.

From the time he was a kid, Harley always wanted to be a farmer. He was born on a wheat ranch east of Wheeler in 1927. He wanted to farm, but not as a dry land farmer because he couldn't take the stress of waiting for it to rain. He started working at 12, mowing lawns and delivering telegrams so he always had some money in his pocket. During that time he knew everyone in town and where they lived.

When he was 14 or 15 years old he started out thinning peaches for Percy Driggs. He used a long stick to strike a branch and knock some of the peaches down. The peach fuzz would come off the peaches, drift down and get under his shirt collar where he was already hot and sweaty. It caused a great deal of itching and discomfort. Percy liked him because he stayed with it and never quit.

Driggs also raised potatoes that were dug with a tractor and mechanical digger. The digger didn't have coulters to cut the vines to keep from plugging up the chain carrying the tubers over and depositing them on top of the ground where they were hand picked into sacks. The only way to get the vines through was to have someone ride on the digger and walk on the vines to pull them loose from the sides and carry on over.

Driggs had a tractor driver who arrived one day so drunk he couldn't even get on the tractor. Harley said, "Percy, I can run that tractor, I watched him, I can do it." Percy said, "Get on and go."

Harley got on the tractor and stayed on it. At 16, he was the operator from then on. Someone else had to tramp the vines through, Harley had graduated. He also planted alfalfa and operated other equipment. He learned a great deal from Driggs because Percy was progressive and tried new practices including fertilization.

Harley went to Eastern Washington College, but quit after two years to help put his sister Irene, and their brother Vic, through high school. After that, he helped Irene through business school. He regretted not finishing college himself; but has never regretted being able to help his family when they needed him.

While stationed at Spokane in the Air Force in 1951, he had an opportunity to buy 160 acres between Ephrata and Quincy. He was still quite young and scared to death with his big purchase at $10 an acre. Other then helping his brother and sister through school, he had saved nearly all he earned from the time he was 12 so he could some day farm on his own.

Harley and his brother, Vic, were both in the Air Force and discharged at the same time in May of 1952. A few days later, the Grant County road supervisor, who knew Harley's family, told him they needed a water truck driver on a project called a "Farm in a Day." He thought that was a great experience to be involved. A drawing was held, the winner got a farm unit. Everyone got together and levelled all the ground, ready to farm with the first water delivery to the Basin. The house, outbuildings and irrigation ditches were all completed in one day. It had great nation wide publicity for the start of the Columbia Basin Project. Harley said, "There were tractors and land levellers all over the fields, people were all over the place, but it was all well coordinated."

In August of 1952, Harley went to Bill Raugust at Raugust Trading Co. in Odessa who was about the only equipment dealer around the area at the time. Mr. Raugust let him buy a tractor and land leveller for a down payment and pay the rest when he could. He had never levelled land before, but he soon learned. It was a mild winter that year and he was able to work right on through.

Harley had just got home from levelling all day at Quincy when his sister begged him to take her and her girlfriend to the dance. He didn't want to go, he was tired, but he had a car and that was the only way they could go. He finally relented and agreed to take them. While at the dance he met this terrific girl, who was a friend of his sister's friend.

Juanita Brown was born at Ellensburg in 1931. Her family homesteaded in Badger Pocket and operated a farm. There were three daughters of which Juanita was the youngest, and a little spoiled. She attended and graduated from Kittitas High School. Her dad let her do everything, she could drive the family car occasionally, but not her sisters, they drove too fast and wild. She was a brunette with blue eyes and a feisty disposition. The family raised potatoes, hay, sweet corn and grain. Her father was going to sell the farm when she was four years old, but she loved that place in the country so much, she told her dad, "You have to sell me with the farm because I'm not leaving."

Her father didn't sell, he kept it until he retired and then moved to Ellensburg. She was an old fashioned country girl, she even plowed for the neighbors as she got older.

On Juanita's 21st birthday, December 28, 1952, she went to a dance in Moses Lake with a friend at Moses Lake's only night spot, Mel's Dine and Dance in Westlake. It was reportedly the wildest place in town. It was where all the young people got together. She was living in Moses Lake and working at a potato shed. It was a night she will aways remember.

Harley didn't want to get married until he had a farm, but something hit him that evening he never experienced before. He didn't have a farm yet, but in May of 1953 they were married. They raised their first crop in Block 71. It was among the first land to receive water and they helped make history. Harley's father had some irrigation experience which really helped make it successful. They had dirt ditches that had occasional cement checks to back the water up near the top of the ditch so it could be siphoned out into the corrugate and down through the field. The checks washed out often and his dad taught them how to place potato sacks around the checks to prevent washouts along with many other helpful hints.

The farming worked out well and in 1954 they bought a 100 acres and built a home in Block 71. Harley and Vic formed a partnership. Vic did the levelling, Harley built the house and their father did the irrigating. The partnership lasted for over 40 years.

The summer of 1953, they rented a house in Moses Lake and drove to the farm. Vic and Harley's dad stayed on the farm in a tent the first year. Juanita came out each day to cook for them on a wood camp stove. She had steak, potatoes and gravy and all kinds of good things every day. All the Ottmars thought she was a fantastic cook, not only that, but she could operate all the farm machinery too.

In June of 1954, their daughter Brenda was born, everyone was excited because girls were scarce in the Ottmar family. The house was up and the farm acreage had increased. Things were going great, should be the end of the story with everyone living happily ever after, however, things weren't quite that simple.

Like all the other new farmers, they had to put up with the wind storms that would howl and move soil and blow out crops. The dust was so thick no one could drive at times because visibility would be zero and the sand piled up along side of the fields deep enough it had to be spread back with a blade. Some of the storms lasted for three or four days at a time. Juanita would wet towels to place under the doors and around the windows to catch part of the sand drifting into the house. "Thank goodness for vacuum cleaners." Harley said. "It was just something you learned to live with."

Harley said, "It was a good life, I wouldn't have given it up for anything. It was fun, it was challenging, it was something I always wanted to do and everything just seemed to go our way."

There were no contracts at first, they had control of the crop planted, planting dates, harvest dates, timing of watering and how much. The only thing they couldn't control was the weather and finances, which made the farm challenging. During that time, insects were not a problem and no sprays were used. The weeds were not thick or so varied as in later years. All weeding was done mechanically or by hand. In 1953, they bought a brand new combine for a little over $7000. That was a lot of cash then.

Through the years they experimented with different crops, such as maize, peanuts, lentils, dry peas, white clover seed, beans and of course, potatoes. The third year of farming they grew the dry peas and when the time came to harvest, they found there were only two swathers in the country, one up on the hill north of Winchester. Harley worked there a full week to be able to use it, but the owner wouldn't let it go that far away. Then they found out Holloways had just purchased a new self propelled Owatona. Harley went to see Earl and Bob. They brought the new swather over and swathed the field of peas that had a terrific vine growth. It was the first peas Holloways had ever cut, but things went well and the pea yield was good.

Life went well until 1959, when they delivered 300 acres of beans to the Mushlitz warehouse in Quincy. Not long after delivery, word got to them the warehouse had closed. They hadn't been paid yet. Every day they called and asked about payment because that was the whole years income. Each call, Mushlitz would say, we don't have the money today.

They finally found the warehouse was shut down because a farmer had dumped 100 pounds of seed treated beans in his truckload of commercial beans going to town and it was dumped unknowingly into the same large bin with all the other farmers' bean crop. It contaminated the entire warehouse of red Mexican beans. They were pleasantly surprised one day when they got a check for the full amount owed. They were one of the few who got paid and felt very fortunate because the bankruptcy court could have called that money back. The court would have found it hard to get because it had been spent quickly on debts.

Harley had an uncle, Rich Ottmar, who convinced Vic and Harley to get into potatoes in 1960, He told them, "Potatoes are always a good price on presidential election years." They only knew one grower in Block 89 who had grown sprinkler irrigated potatoes; that was Adrian VanHoltz. They talked to VanHoltz about his experience, got an old planter from Ellensburg and planted their first crop of potatoes under wheel line sprinklers. They raised a great crop just as they hoped, with a good yield. The tubers were big and the price was terrific. They made enough money to buy new spud equipment for the next year and increase the acreage.

They worked with Quincy Farm Chemicals, run by Tony Romano. Tony had furnished the fertilizer and was vitally interested in the spud crop quality and volume. He could see the future potential for potatoes in the area. The bigger the potato, the better the price at that time, and they could stack that first crop in their arms like cordwood. George Yoshino of Quincy ran all the Ottmar potatoes that year and the following years through his fresh pack shed. They tried to get the first potatoes in the warehouse every year because the first ones brought a premium price.

In 1963, they had a crop of early potatoes. It was still early spring, the plants were up and there was a frost warning. Young potato plants could be damaged by a frost. Harley and Vic set out thermometers in the field and arranged for a helicopter to be there early the next morning. The copter cost $100 an hour, which was a lot of money then, but there was no way they were going to let those vines freeze.

As the sun came up in the morning, the temperature started to drop. Harley started the copter moving back and forth across the field moving the air. Where the copter passed, the temperature came up two to three degrees. There were 80 acres of potatoes, but the copter stirred a wide path going back and forth. The temperature never quite got to freezing, and the crop made it through. Innovative people meet the challenges that come their way.

That fall they had Cliff Boorman dig and haul the spud crop in order to keep the shed running at capacity. It was a good harvest with no big problems. All the crop was hauled, sorted, shipped and delivered to markets. Meanwhile, a shipment of potatoes in New York was discovered with traces of Dieldrin.

Dieldrin was used for wireworm control in the Columbia Basin area by some growers. It was approved by the USDA and Ottmars had applied it according to the label for usage. The USDA inspectors came swarming into the area looking for any trace of misusage or other potatoes to which the chemical had been applied, but the tubers were all gone and there was nothing to find. The inspectors were on the farm every day for a month trying to find out what happened. They measured out the acres planted, the amount of chemical used, but could find nothing out of order. Since there were no tubers left and all use requirements had been met, they finally gave up and left. It brought out another hazard of the industry in which they were engaged. So as Harley said, "It made life interesting, and worrisome at times."

Boorman dug potatoes for them a number of years and then Holloways after that. Earl and Ruby were always right there on the job. One harvest was in a bad price year and really set them back financially. It took a few years to pay things off and they had to borrow operating money as well.

Many people weren't as fortunate as the Ottmars. Lots of farmers went out of business for different reasons. There were those who had bad luck, those who didn't gain their experience quickly enough and some who were not willing to put forth the extra effort in order to make it through the hard times which were considerable.

One of their early neighbors was John and Shirley Morris. They became close friends over the years. One of their boys, Paul, used to wade a half mile through the snow when he was five and six years old to play with Harley and Juanita's daughter, Brenda. He would usually bring a note, reading, "Can I play with Brenda?"

They sold the home place in 1960 and moved back to Moses Lake. The farming continued, but more of it in Block 89, closer to town. Harley was financing with Production Credit Association (PCA) for money to farm. In 1957, Ralph Neal, manager of the Yakima Valley PCA asked Harley to work for them and set up an office in Quincy. Harley told him, "No, I'm having too much fun farming, I don't want to do that."

Mr. Neal would ask him off and on about going to work for them. After the move back to Moses Lake a few years later, he decided he wasn't doing anything during the winter months of 1962 and 63, so he worked for them part time and decided he liked it. PCA wanted him to work full time. After considering the proposal, Harley decided to accept the position and Vic would continue the farming.

Harley helped in the evenings, weekends and vacation times. He ended up with two full time jobs; at least he didn't get bored. Harley liked to work, he liked people and thought he understood them. He was happy with what he was doing and had a feeling of accomplishment. He enjoyed working with farmers starting out with nothing and seeing them grow mentally and financially to become solid productive citizens.

He can still think of many farmers who are quite well-to-do in the 1990's that started with nothing but a wife, an old car, a tractor, a plow and the determination to succeed. He felt to be able to help others was not only right, but it was fun and fulfilling. Banking was different in those times. The loan officer could judge the type of people wanting loans. If they were hard working, willing to learn, able to cope with adversity, and had good character, they had all the things it took to succeed in a tough occupation. The loans were not approved strictly upon a financial net worth and bravado. It seems Harley had a knack of picking the winners from the losers in the game of farming.

Harley said, if a person came in the door with a big cigar and a big fat belly, he was dead on arrival. To make it as a farmer, you were lean and mean, and you had to work.

There are people all over the Basin today that got their start with PCA when everyone else had turned them down for loans. It was because of the policy and philosophy of the organization during those years that made it possible, along with competent personnel such as Harley Ottmar.

About 1967, the Yoshino potato warehouse got into financial trouble. Ottmar's didn't get their last check on the year's crop. They were fortunate, however, because many growers got very little of the amount due them. Ottmar's crop was delivered early and Harley was there each month to collect; by the time the payments stopped, they were close to being paid out. Was it good luck, or was it good management?

In 1992, after 40 years, Vic and Harley quit farming and retired to enjoy other things in life other than work. Harley and Juanita have a great time with their grandchildren. Among other things, they instilled a work ethic in them through truck gardening and lawn mowing. They learned how to grow crops and market the crops they raised. They taught the children the same values that kept themselves in good stead for so many years. Harley and Juanita had a feeling of fullfillment as the children became older, a reflection of what was learned in the truck garden could be seen in their development as adults.

They both belong and enjoy a Horseless Carriage Club for old cars. Harley was elected president in 1999 and as always, used his and Juanita's talents and energy to make it a better and more active club than ever before. They have restored at least two classic cars of their own, a 1964 ton and a half Ford, and a 1950 two door hard top Cadillac. Their car trophys are stacking up.

Harley said, "I wouldn't trade my working life for anything, I would do it all over again." "As far as the Basin goes, it was very good to us. I'm sure glad I don't have to go out there and ride those tractors anymore. If you worked hard and managed your time, marketing and money properly, it was good to a lot of people."

"Farming is the only business where you create wealth for the nation out of nothing and it's renewable. I think farming is America's greatest resource, you can live without iron and material things, but you can't live without food."

Harley and Juanita have always enjoyed whatever they were doing whether it was farming, working with the grandchildren or helping on a special project like antique cars. They didn't go through life for the end purpose of getting rich, they simply loved what they were doing and proud to have filled the role they have in agriculture and the development of the Columbia Basin and their family.

Leveling ground for a farm

Chris, Harley and Vic Ottmar
1956

Harvey out standing in his field

Planting a bean crop

Ike and Alice Parker

Love and Preseverance

It was cold, not enough to freeze the rain that was falling, but enough to chill a slim, dark haired boy riding a large roan colored horse. Although the roan had long legs it was difficult to break through snow that came almost to its chest as he drove a herd of range cattle toward home. Home would mean feed for the cattle and shelter and a hot stove for the cowboy. It was only one day of many in a life that would challenge the brown eyed young rider.

It was the day before election day in November, 1946. There was no wind, just snow, 40 inches deep on the level. It was early for snow and cattle were still out on the open range. Nearly all the cattle were stranded away from a source of feed. The first day after the storm, A.E. Parker and his 14-year-old son Ivan rode horses seven miles to bring in a few head. Ivan, whose nickname was "Ike", rode a work horse to break the trail through the deep snow cover. It took the whole day to bring the small bunch of cattle back to the ranch.

There had been no votes cast as election day passed, no one could get to the polling place. All the roads were still closed. Two weeks after the storm had hit, A. E. and Ike left on horseback again, this time for a neighbor's ranch where most of their cattle were located. They rode all day to cover the 11 miles, breaking a trail through the snow to the ranch where they were put up for the night. The snow was still more than three feet deep as they rounded up the cattle the next morning. The first job was to separate theirs from the neighbor's stock. When they completed separating the herd, they started the push for home. The snow was bad enough to work in, but then it started raining. It was a cold rain, not cold enough to freeze into ice, but it was wet, their clothes were wet. Despite the rain gear, it seeped down their necks and up their sleeves and water ran into their boots. The movement was slow and they almost froze before reaching another ranch seven miles away, late in the day. The rancher saw them coming and met them at the barn to put the horses up and feed them. Meanwhile he sent the father and son to the house to thaw out. They were stiff with the cold, their clothes were wet and they wondered if they would ever get warm. Right then, that's all they wanted.

That same day, about noon, they gathered the cattle next to a creek where they needed to cross a bridge. There were cracks in the bridge and the cattle were leery of it and hard to push across. A.E. kept telling Ike to, "Watch it, watch it, you're going to ride your horse off into a hole and get hurt." Ironically, it was A.E. whose horse did exactly that. He didn't say a word about it to his son, even though his testicles were smashed against the saddle horn as the horse had suddenly lurched before regaining its footing. He didn't tell his son he was injured, he just gritted his teeth, hung onto the saddle horn and endured the pain.

Inspite of the injury and pain, the third day they were able to get the cattle to their home place behind the windbreak where they had feed and shelter. At least it didn't rain that day. Ike's mother had no way of knowing how they were doing, it had been three days with no word. They didn't have telephones and with the roads still closed, no one had come by. Ike's father was badly hurt and the only way out was by airplane. The next morning, Ike took the biggest horse they had and rode nine miles to the nearest telephone to call for aid. A ski plane was 40 miles away and able to land a quarter mile from the house. He was transferred to the nearest hospital.

Meanwhile Ike had a list of groceries his mother sent with him to pick up in town where he used the phone. He rested at a relative's home while they filled the grocery list. When he got the supplies, he mounted up and rode the nine miles back home the same day. Ike and his brother tended the cattle while their father was recovering. They had to shovel the shocks of feed from under the snow and load it on a horse drawn sled and haul to the stock. The snow had come so early no one was ready for it. The feed wasn't hauled into the stack yards yet. Hay drops from cargo planes were organized to keep cattle from starving. It was a hard winter, but it made a boy a man in a hurry.

Ivan was born at the family home near Karval, Colorado, in 1932. He was third of four sons by A.E. and Hazel Parker. His oldest brother was kicked by a horse shortly after Ivan was born. The only medical help was an animal veterinary in the area. By the time they were able to get to a medical doctor the brother died. The cause of death turned out not from being kicked, but from appendicitis. The dryland farm raised 500 acres wheat, milo and corn to feed primarily their own cattle. The rest of the 20 sections of ground was livestock range. His father waited until tractors had rubber tires before getting his first one, using horses until then. They sold calves at weaning age, a top of 400 pounds compared to modern methods of 600.

Ike's father moved to Colorado from Kentucky in a covered wagon when he was a young boy. Some of the places they tried to settle was Lahunna and Hugo, Colorado, before finally arriving at Karval leading a milk cow behind the wagon. There was butter on the milk every night when they camped. Having hung on the wagon all day would churn it nicely. One of the items on the wagon was a family keepsake, a handcrafted clock with chimes. The clock is now over 200 years old. Ike now owns and treasures it.

His folks arrived in Colorado in the late 20s. They started farming but it was tough going. His dad stored a crop of pinto beans in the loft of the barn which he couldn't even give away. The word was out you could get rich in California so in 1932, in the middle of the depression, they packed up to go further west. When they arrived in California they, along with thousands of others, discovered it was not the promised land and no jobs available there either. A year after leaving, the family returned to Karval. The four ton crop of beans were still in the barn and the price had jumped to $30.00 a hundred. It was enough to get them started farming again.

When Ivan's father went to the court house upon buying his first half section of ground, he didn't have the fifty cents it took to pay the abstract stamps at the title company. The clerk dug in his own pocket and paid it for him. Over a period of time he acquired 10,000 acres and nearly all at fifty cents an acre. Most of it was range land. Everyone in that area raised feed crops on selected pieces of dryland to feed the range cattle during the winters. Normally the income was derived from sale of livestock.

When Ike started grade school the first year the area had busses for students; he went to a one room country school for all eight grades. During World War II, Karval didn't have a high school so when he was old enough, Ike attended the school at Hugo. The students boarded in private homes for the school year. Two widow ladies managed the homes which were like dormitories, one for girls and one for boys. Ike was not one to like school and instead of finishing high school, went to work on the family farm. His parents liked to travel and Ike took care of the farm and ranch work. It wasn't long before he become attracted to a pretty blue eyed country girl he had known all of his life. He and Alice were married in 1952.

Alice was the daughter of Ralph and Sara Taylor. They moved to Karval in the late 1920s from Kansas. She was born in a midwife's house in Hugo in 1933. Alice had three sisters and two brothers. It was during the depression years and her folks had a small grocery store on the corner of their farm. Her mother ran the store each day and Alice began her social life at an early age. A crib was placed in the store for Alice to spend her time while her mom waited on customers. All the young children would end up in the crib to play and sleep while their parents did the shopping. Alice said, "Ike and I slept together 19 years before we got married."

Alice also attended a country school, but different than where Ike went. Her school had three grades per room. After grade school she also boarded at Karval until she graduated from high school. Alice and Ike grew up together, went to 4H and high school together. She said it really irritated her, for years while in high school, they helped each other with their 4H steers, washing, grooming and preparing them to show. When they went to the county fair his steer always won.

The closest movie theater was at Hugo, 40 miles away so their dates consisted of working with the cattle or Ike would help Alice with her chores. After the chores were completed they played cards and games. Every other Saturday night a dance was held in an old army barracks that had been moved onto the corner of Alice's dad's property. They were old country dances, all the family went, little kids through grandparents. Benches were along side the dance floor where people could watch and visit and the kids would eventually fall asleep. The smaller kids went to the kitchen to dance so they didn't get trampled by the big folks.

Alice went to Fort Collins to college after high school graduation. She had surgery after two months in school. The doctor advised her not to go back to her classes until she recovered her strength. Instead of going back to college, she said she got her MRS degree on April 19, 1952, that's short for Mrs. Ivan Parker. Alice had a sister that married Ike's brother.

Ike was on the verge of going into the military because the Korean War had started. He had his physical completed and was just a few minutes from signing at the draft board one day only to be advised he could do more good for the war effort at home farming than going into the military service. Alice was pregnant with their first child and the chairman of the draft board, Mr. Love said, "No, you are not going, no way."

Ike worked with a D-6 and a D-8 Cat for a year and a half after they were married. He built small reservoirs used for stock watering ponds during the dry months. They lived in a 17 foot trailer house for the first six months. They moved it to a different location every week following his construction jobs. The Cat and operator were hired out at a price of $10.00 an hour. If they pulled a scraper, the charge was $12.50. Ike was paid one dollar per hour and they thought they were doing great. They had milk cows and chickens, so on the trips to town, cream was sold to the creamery for a small amount and eggs were traded for groceries.

Ike and another young fellow had to move the D-6 and D-8 Cats from one job to another, sometimes 10 miles in a day. The boss wouldn't let them run it any faster than 5th gear and when they moved it was in 3rd, which was slow and tedious. One hot day going across range land they headed the cat in the right direction, climbed off the back, got in the more comfortable and quieter pickup and drove along side until the Cat started veering to one side. Then one of them would crawl over the back, straighten it out and go back to the pickup again. This process was working quite well until they got the pickup stuck and the Cat still chugged on its merry way. It took Ike's friend three-quarters of a mile running before he finally caught it.

Two years later, 1964, Ike and Alice started farming on 720 acres of their own. They acquired a place next to Ike's folks' ground which turned out to be worthless for farming. It was sandy. Ike said "It was terrible, we could never grow anything the ten years we were there." He still worked construction to pay the farming bills. When the creek was running they had water. During the dry years, the creek dried up and they hauled water for everything including livestock.

In 1959, A.E. and Hazel, Ike's folks, drove through the Columbia Basin on their way to Alaska. Alice had some cousins who moved to the Basin after drawing some ground and the Parkers stopped to visit. After seeing the Basin, which was in the middle of its development, they thought it was a new world of opportunity. They returned to Colorado with glowing reports of the area and felt Ike and Alice should take a look.

Ike's dad drove to Washington with them in 1964 to look the farms over. When it was time to go home there were reports of a big storm moving in. A. E. was quite concerned about the storm and thought they should leave quickly as possible. Their plans were to leave early the next morning. They drove to Baker, Oregon and stopped for the night. Alice went to bed, but in the next room they could hear an anxious A. E. pacing back and forth at midnight and still at one in the morning. Ike and Alice decided they weren't going to get to sleep anyway, so they got up and left at 2 AM. Satisfied that they were all getting away from the impending storm, Ike's dad lay down in the back seat, went to sleep and didn't wake up until they consumed a complete tank of gas.

They put their name in for the land drawings. In 1962 their name was drawn and they made a trip to see the ground. They didn't like it and returned home. When Block 81 was developed in 1965, they returned and purchased their first ground. It consisted of 226 acres from four different private owners and was in two units of Block 81. They thought, "What have we got to lose? If we lose everything and can't make it, we can go to work for wages."

There was a farewell party for Ike and Alice when they left Colorado. At the party, an older German man they knew told them, "I'll give you one year and you'll be back." Right then they decided, they were not coming back. They were bound and determined they were going to make it. It was the kids that the move was hard on, having to change friends and especially leaving their grandparents who were a big part of their lives.

The first year there was no income. The water didn't arrive until October, just in time to get a few newly leveled fields wet. Ike's dad hauled manure from a feedlot south of Quincy almost 20 miles away and covered the fields to prevent blowing sand. That was an accomplishment in itself. He used a single axle truck which made every trip, and climbed the Frenchman Hill grade on Adams Road.

They had a lot of company the first year and with very little income, they didn't know how they were going to feed everyone from one meal to the next. They heard about food stamps and when they asked, were informed they owned ground and didn't qualify. Alice worked at Chef Ready, a potato plant in Othello part time and Ike had brought his combine from Colorado and was able to do a little custom work. Alice thought she should find a full time job, but Ike said, "No, we would have to hire a man and pay him more than you could earn off the farm." So Alice said, "I was the hired man, along with our kids, because they really helped us."

The Parkers had one son, Perry, and two daughters, Susie and Carla. Perry was 12 years old when they arrived in the Basin. He had been driving tractor since he was 10 years old and he and Alice were doing much of the farming at first. Alice said at that age he could tell when things were going wrong sooner than she could.

Their hardest period was the early years in the Basin with winds blowing the light sand, and trying to hold the sandy, corrodible head ditches together. There were times while setting water Ike and Alice couldn't see each other 30 feet apart because of blowing sand. Ike thought they were behind on watering once, so he had ordered seven foot of water for the fields, which meant a lot of siphon tubes to set quickly. That was also the morning the hired man notified them he was quitting. But they never regretted coming and felt they could meet the challenge and conquer whatever happened.

Through the years they expanded by purchasing some nearby acreage from Tom Yamamoto in 1972. Ike tried to rent Dr. Piper's ground next to them. John Baird, an attorney from Ephrata, was managing the property. Mr. Baird declined when Parkers wanted to rent, because he wanted a large farmer with lots of financial backing to lease the ground instead of a small farmer. The doctor's farm was rented to a large potato farmer from Warden. The farm wasn't handled correctly and there were no potatoes harvested from it. The new tenants, because of bad farming practices, had it blowing so badly it deposited tons of sand on top of Parker's land that was downwind. Finally Dr. Piper rented to Ike for paying the water charges for a year and then the next few years for paying water and taxes.

Their first home was a 10 x 55 foot mobile home. They placed it on the corner of the farm. The fine sand sifted and drifted every time the wind blew from the newly developing ground west of them. Every morning there would be a white spot on the pillows where their heads had been. They had to haul their drinking and cooking water. Their baths were in a nearby irrigation canal. They mounted a water tank on a truck. The tank was filled in Royal City, than parked on high ground near the trailer. A hose was run from the tank to the mobile and used gravity flow to furnish their needs. Alice said they had to be careful taking a shower to be certain the water didn't run out after you got all lathered up and not be able to rinse it off, the soap got pretty itchy.

They went to their banker about a loan to build a shop. The banker told them, "No, we're not going to loan you money to build a shop. But we will loan you the money to build a shop and a house." They tried to buy 60 acres on the hill above the farm from an absentee owner several times before with no success for a future home site.

It was a fantastic location for a home and with the new loan, they thought it worth one more try. They went to Royal City to use a phone to call and before that day was over, they had a new home site and the money for a shop and home.

They hired two carpenters, who with the help of Ike's brother, built their home in 1968. It is located close to the top of the Frenchman Hills north west of Royal City. A domestic well was drilled near the new shop. The well developed a problem, the water was flowing on through and not coming high enough to utilize correctly. Ike thought he would cure that, and dumped a yard of cement in the bottom. The well was still losing water. Ike thought to himself, "I'll fix you this time". He did! After dumping in three more yards, it filled clear to the top. They had to drill a new one, which is still in operation.

The view from their living room has a panoramic view of the Royal Slope area. It's like being suspended high in the air looking down on the different shaped fields and varied colors of growing crops. They appreciated the new home and so did the neighbors. One evening not long after the house was completed, Sax and Arlene Fleming showed up with all their kids and sleeping bags. The chemical Thimet had been sprayed on the fields next to their house. They couldn't stand the odor and came to the Parkers for the night. Sleeping bags were scattered all over the living room floor. "Those were good times, people felt comfortable doing that", said Alice.

The crops they raised were alfalfa, wheat, dry peas, red Mexican and pinto beans and sugar beets. The sugar factory closed down in the late 1970s to end their sugar beet production. Later they had some sweet and field corn and green peas. Most of the ground was nearby until they decided to expand into the Black-Sands area south of I-90. They farmed there two years. The first year the wheat shrivelled dramatically and as a result the price was cut drastically. The second year they had two circles of beans. "When we planted they were bringing $30.00 a bag, the crop was good, but the price fell to eight dollars. We really got an education in the Black Sands", Ike said.

Periodically times were tough financially; one year they tried cutting costs by not weeding the beans. Alice said, "We practically lost the whole thing because the night-shade was so thick. We tried to save money and it cost us money, big time. We learned you have to spend money to make money." Ike said, "There was a period of time when every piece of machinery I had on the place was borrowed from a neighbor. That was just part of life here in those days. We've had wonderful neighbors here, and we couldn't have made it without them."

After the Mt. St. Helens eruption and the ash falling three to four inches deep over the ground, it was a whole different ball game. "We had to handle our ground differently. Before the ash fall, it was hard to get the ground wet, but afterward, the ash seemed to change the texture and moisture holding capacity of the soil. The day the mountain blew, they were planting beans. Tim Snead was on the only tractor with a radio. When the Parkers saw the dark clouds approaching and could see lightning and hear thunder, they thought a really bad rainstorm was coming. Alice and Ike were changing water and decided they had better hurry or they were going to get awfully wet. They stopped at the tractor to talk with Tim and he told them, "Oh no, that's not a rain storm, that's part of Mt. St. Helens coming."

Upon getting the news, the Parkers headed for the house and Tim got in his car and headed for Moses Lake trying to beat the ash. Tim was driving a little sports car and was driving 110 miles per hour all the way to Moses Lake but he couldn't outrun it. It ruined the car engine. Alice said, "It got so dark by 2:00 PM, we couldn't see the railing eight feet from our glass doorway." When they opened the door the air had a smell of sulfur. The cattle in the pasture were bawling continuously. The Parkers wondered if the cattle would be alive the next morning. Birds were hitting the window, trying to get in where there was light.

The alfalfa was covered with the ash. They ran excess water through the sprinklers to wash it clean enough to swath. Afterward the bales had so much ash in them the hay was not saleable, so it was burned. They measured a foot square on the pickup and weighed it dry. They did the calculations and estimated the weight per acre was 92 ton.

Over the years Alice said, "We've had some tough times, but we've also had some good times too. We feel its been worth our time to move here." They cleared and developed most of their own ground. Their farm was the first to be leveled with the new 619 Caterpillar self-loading scrapper. Ike, being an experienced operator, did his own work. They also used a Case 4-wheel drive pulling an 8 yard self loading scrapper. They used gravity flow irrigation for years before finally converting to center pivot systems.

Ike had his second cancer surgery in 1982. Their son Perry was working cattle horseback and was thrown off and ended up with a broken ankle. Alice had extra responsibility and work along with the mental stress of the family. It was during this time some women friends got her to take golf lessons at Desert Aire. She said, "That was the best medicine I could have had. It took my mind off what was going on at home and I got so angry at that little white ball that it relieved the stress." The women then got their husbands involved on Sunday afternoons. It got old friends back together that had drifted apart after their children's school and 4-H activities had ceased. "It brought us back to being neighbors again," said Alice.

They have enjoyed playing golf as a hobby, but they have been active in many groups and organizations since coming to the Basin. They were involved with their children in 4H and part of that time as 4H leaders. Alice was on the State Fair Board for six years before becoming active in WIFE (Women Involved in Farm Economics). She served as national president of WIFE in 1988-1989. There were 19 trips to Washington DC while president. She was always promoting agriculture through lobbying efforts and testifying on issues.

She also served on the Washington State AG Council where she served as president and on the Grant County Fair Board for 18 years. She was a founding member of the AgFarmation committee that erected the large electronic billboard along the freeway at George, Washington. She was finally offered a paying job as Executive Secretary of the Columbia Basin Development League, which she accepted. Alice is probably the best known spokeswoman for agriculture in the Basin. Ike, knowing the importance of people speaking up for the farming industry, has been supportive and helped her in those endeavours as well as being a member of the Cattleman's Association and Farm Bureau.

Alice said, "We've been in agriculture all our lives and see the importance of it to the country. I think we've stepped back for too long, thinking everyone cared about what was happening to agriculture, but they've totally forgotten it. We've got to start speaking up really loud. If the trend continues, we will be dependent on third world countries for our food, and I just don't want to see that." Ike added, "Just a little common sense would help."

Ike was diagnosed with prostate cancer in 1979. It struck again in 1982 and advanced into the bone. Since that time its been a constant battle. By year 2000, he's had over 100 radiation treatments and 37 bone scans. Ike is a firm believer in a positive attitude. He also credits a certain medicine he has used for many years. That particular drug was discontinued in the United States and all of a sudden it was not available to him. Without the use of it for two years the cancer renewed its attack. They finally located some in Canada, and through the Internet Alice found a source in New Jersey. With resumption of the drug, Ike feels the cancer is under control again.

Their son lives in Moses Lake and works for Wilbur Ellis out of Quincy. Wilbur Ellis bought the operation from Nexus AG Chemicals in the spring of 2000. One daughter lives in Spokane and the other has a sewing business in Great Falls, Montana. Ike and Alice have 10 grandchildren of whom they are extremely proud. The oldest is now serving in the U.S. Navy aboard a Trident submarine, the rest are near the top of their classes in school.

Living high on the hill has its ups and downs, especially in the winter. When snow is on the ground the snowmobiles appear. People sled down the long slope and the snowmobiles pull them back up. One day Alice and Ike fed 40 people as they kept coming in to get warm in between sled runs. Alice normally keeps a large batch of cooked beans in the freezer along with hamburger, so during those types of days, chili is on the menu. Ike and Alice are the kind of people who like their friends and neighbors dropping in.

One of their neighbors, Mike Brown, asked them, "When are we going to have a party?" Alice said to set a date. She was WIFE president at the time, the date they set was two days after she got home from a Washington, D.C. trip. When she left, the invitation list was for 30 couples. The party was on Sunday, and when she arrived home Friday night the list had grown to over 100 couples and family. One invitee called to ask if family meant all her kids and grandkids? Many of the guests brought food and helped set up tables and everyone brought salads or desserts including ice cream. Marje Brown went to Hermiston and brought a pickup load of watermelon. Ike and Alice provided the beans, potato salad and the meat. More than 300 people came. Alice thought it was so nice because there were three generations present, truly a family affair.

Much like Ike, as a 14-year-old boy enduring the snow and cold, the Parkers have continued to meet and master the challenges that have come their way. Ike and Alice lived through the dry conditions in Colorado, the blowing sand in Washington, the ash of Mt. St. Helens, financial hurdles, and the bouts of cancer. They, like so many of their neighbors, have worked, persisted and endured to overcome many hardships and trials, not only that, they had fun while doing it.

Ike's father, A.E. and crew in
Colorado, dipping Cattle for lice

Sue, Perry, Alice and Karla at home in Colorado

Ike and Alice on their Wedding Day, April 19, 1952

The Parker's first home in Washington Royal City in 1965

Message on the counter top after a dust storm. 1976

The snow of November 2, 1946. Ivan in the center. Colorado

Larry and Eleanor Richardson

Love and Preseverance

The Richardsons, like so many other people in the Basin, had to work long and hard for what they achieved after moving to Quincy. They also exemplify a family who knows there is much more to life than hard work and financial gain. The love of family, religious belief and friends have helped guide them through the years. An example of their feelings have been expressed with a few words in Christmas cards they have sent in the past, "Thank you for sharing your life with us."

Larry was born at home in Latah, Washington, in 1921. His family operated a dryland wheat and pea farm not far from the Washington-Idaho border. He had one brother and one sister. He attended the same one room school where his mother had taught years before. Larry went to Latah and Fairfield High Schools before choosing Gonzaga University in Spokane to gain his scholastic knowledge and wisdom. He joined the U.S. Army Air Force in 1940 and was sent to March Field in California for basic training. Larry called it boot camp instead of basic training because it was boots, boots, boots. The base covered thousands of acres and he said he covered every square foot of them.

The army thought he would do well in communications. After basic training, they shipped him to Illinois by train. He arrived in October, the wind was blowing hard and it was cold when they arrived at the base.

"Where is the barracks?" Larry asked. The Sergeant pointed off to one side and said, "right over there." "That looks like tents" said Larry. "That's your barracks," said the Sergeant.

The tents had small coal fired heaters that looked like an upside down funnel. The stove pipe went up through the top of the tent. Invariably someone would accidentally knock the pipe down which resulted in soot being scattered all over the beds and everything else. Larry said he never got warm all the time he was there.

When he shipped out from Illinois he went on furlough and through mutual friends, he met a gorgeous young lady in Spokane by the name of Eleanor Mauk. His next stop was Pearson Field Washington, directly across the Columbia River from what is now Portland International Airport. The thing that impressed him there was an aircraft used for observation. It was a very light single engine plane called a 047. It was capable of flying at a very slow speed. Larry enjoyed watching it land. If the wind was blowing, the ground crew would run out on the runway and grab the wings to pull it down to the surface. It was so manuverable it could almost make a square corner when turning. They designed it to fly slow, low, and take pictures.

When Larry received orders for his next move, it was to Hamilton Air Base north of San Francisco. At Hamilton, they assigned him to Borneo. But before shipping out, war was declared and his orders changed. The new destination was Christmas Island where he was to build and operate radio stations. His tour also took him to Honolulu, Fiji, New Guinea, and New Caledonia.

Eleanor Mauk was born at home in Spokane. She had one younger brother. Her father drove street cars and busses in the area. He was elected president of the Central Labor Council and served for 20 years. During his tenure, he was called upon to meet with President Truman and later, John F. Kennedy.

When she got out of high school, Eleanor had an interview with KHQ radio in Spokane. She did so well they encouraged her to go into radio, but she declined, to her later regret.

While working as a candy girl at Woolworth Department Store, she attended Kinman Business College at night. She then worked at the Spokane County Courthouse. Eleanor wrote to Larry in 1942 and asked what he thought of her joining the military. She also asked her boss the same thing, then joined before either one gave her an answer. Seattle was the closest place to enlist. She visited an aunt there in December of 1942 and become a U.S. Navy WAVE.

Boot camp for her was in the New York Bronx. She had never seen a skyscraper before the train arrived at Grand Cental Station. She still didn't see the sights because she caught the subway straight to camp. It was six weeks before Eleanor got a day pass and along with new friends, she rode the subway back into New York for the sights and lunch at the Astor Hotel. After some schooling in Georgia, they assigned her to Washington, D.C. as a Communications Specialist Q. While stuck in Washington D.C., she passed the test for Yeoman 1st Class and asked to be transferred to a radio station on Bainbridge Island in Washington State. She wanted to be closer to home in case Larry was able to return to the States.

Pearl Harbor was struck a terrible blow by the Japanese in a surprise attack on December 7, 1941. Before the day ended, 18 ships were sunk or badly damaged, 188 planes destroyed and 159 were damaged. More than 2,400 Americans were killed and 1,178 wounded. Larry and Eleanor were among 15 million men and women in the military service by the end of the war.

In 1944, after three years in the Pacific, Larry was granted a 30-day furlough if he promised to return for the invasion of the Philippines. Eleanor couldn't get a leave to go home, so Larry's parents allowed him to fly to D.C. and spend part of the time with her. After a short time to get reacquainted, they were married in Rockville, Maryland. A week later the war parted them again and Larry kept his promise to return to the South Seas.

In the Philippines his assignment was at Zamboanga on the Island of Mindanao. (Some may remember the song, "Zamboanga where the monkeys have no tails".) Of course all mail was censored during the war so Eleanor and Larry's family NEVER knew where he was. His last duty was on the Philippine Island of Leyte, and then a 30 day trip home by boat where he contracted yellow jaundice and lost 40 pounds in two weeks. Eleanor finally got to him in the hospital on Angel Island on his return. The boat stopped at Alcatraz every day on the way to Angel Island and one day they claimed a prisoner escaped and was onboard.

On V-J Day, Larry was on the way home, the war was over. The navy was lenient to a married WAVE. The two were able to go to Texas for his discharge. However Larry had to ferry back and forth to Bainbridge for a few weeks before Eleanor received her release. There are three honorable ways to get a discharge—if you had enough service points, if you were married, and if you were pregnant. By that time Eleanor had all three.

United States officials were convinced that a half million Americans would die in an invasion of Japan. Hoping to avoid such loss of life, President Harry Truman ordered the atomic bomb dropped on Hiroshima on August 6, 1945. Three days later, Nagasaki was the target of a second Atomic blast. It was only five days later on August 14, hostilities ended. The bitter war was over after three years, eight months and 26 days.

Almost 35 years later, Larry decided to show Eleanor some of the places he had been. Along with their son Doug and Larry's brother Bob and wife Carole and daughter Annette, they headed for Fiji. As they crossed the international dateline, they missed their wedding anniversary for that year, but Doug had champagne delivered to them by the steward at 3 a.m. aboard the airliner. From Fiji, they went on to Espirito Santos in the New Hebrides Islands

In the jungles of Espirito, Eleanor, with a big frond in hand was trying to avoid big yellow spiders while Larry, with a machete flailing at the jungle growth, took off trying to find his old encampment. She wondered if she would ever see him again, but he found the spot where his tent had been many years before. Larry said when he was stationed there, they were careful when getting anything out of a box because cobras liked small spaces to hide. Most of the encampments had been completely covered over by the vegetation and was unrecognizable.

They stayed at the only motel, as a matter of fact it was December and they were the first guests since the previous February. It was there they talked to three fellows who were divers. Their purpose for being there was to explore a sunken ship and check for any ruptured fuel tanks that might pollute the waters. The ship was located between two islands in a narrow channel that had been mined during the war. The ships sailing through the channel needed to be guided to miss the mine field.

During the war, Larry and another serviceman were on that island at the time the same ship was sunk. They were on a hill above the channel one day as an ocean liner started through the channel. They noticed the ship suddenly shudder, then slowly roll on its side, it had hit a mine. The picture of the ship laying on its side appeared on the front cover of Life Magazine. The picture showed the sailors scrambling over the side trying to get off the vessel. The caption under the picture said they had encountered an enemy mine.

"Phooey," said Larry, "they hit one of our own mines, only because the Captain wouldn't wait for a pilot to come out and guide him through. The worst of it was, that ship was loaded with beer."

The island had contained three bomber and three fighter airstrips. Since the troops had left in the late 1940's, the area had been virtually untouched. They had saved only one strip for commercial use, the rest had disappeared under jungle foliage.

While the divers were in the area, they heard about the Million Dollar Point. It was not far from the ship they were to inspect. The reason for the name "Million Dollar Point" was an incident after the war ended. That particular island had a jetty of land that extended into the water where the ocean floor was very deep. An air strip was nearby and a multitude of tanks, cranes, jeeps, trucks and huge metal buildings and other heavy duty construction equipment was still there.

The orders were to evacuate the island since the war had concluded. U.S. Officers offered millions of dollars worth of equipment to the French for $50,000, a fantastic bargain. The French, thinking they could get it all for nothing when the troops moved out, refused the offer.

The General in charge, decided not to let all that equipment go for the taking after they pulled out. He ordered everything run off the point and into the deep water. The inventory of the whole supply depot went to the bottom. They tore down and dumped huge metal warehouses into the water. Larry said there was not a scrap of anything left when they finished. The divers told him, it was the strangest sight you ever saw, large piles of tanks sitting on top of graders and trucks, etc. Therefore the name, Million Dollar Point.

A year or two later, the two Richardson brothers and their wives were off to the Philippines via Japan, a long, long trip. Somehow their reservations in Manila were fouled up and they were given the Presidential Suite. They really didn't need the large conference room, but it was a lovely place. A few days in Manilla and they proceeded to Mindanao and the village of Zamboanga. A flooded creek prevented them from reaching the actual site where Larry had lived and worked, but everything about the area was so picturesque, it was hard to believe a war had taken place there. It was interesting to see Japanese tourists taking pictures with their cameras of a land their ancestors had so badly mutilated and cruelly killed the inhabitants.

Both Larry and Eleanor were discharged from the military in 1945. Larry resumed school in the winters and worked on the family farm during the summer until he graduated from Gonzaga. While he was finishing school, their daughter, Laura, was born. Eleanor worked as a proof reader for a college magazine to help with expenses. They drove through the Basin on a windy day, dust and tumbleweeds were flying across the road. When they passed Moses Lake, Eleanor noticed the green fields and long rows of poplar trees just out of town. "It's pretty and they are doing a lot here, but I would never want to live here." she said.

After graduation Larry found that farming was a tough way to make a living. Through a friend he managed to land a job with an International dealer in Spokane called Rental Equipment that sold large construction machinery. The summer of 1959, he hauled a forklift to what he thought was, a God forsaken hole. He said it must have been July because dry peas were being harvested. There wasn't any freeway then and trucks had no air conditioning. It was hot travelling the narrow highway across dryland wheat country through Connell and into the Quincy Basin. After unloading, he stopped at the Turf Café in Quincy to get a coke and cool off. Someone in the café told him it was 110 degrees F. "People don't live in this kind of heat, I'm getting in my truck and I'm never coming back", said Larry.

Larry was offered a job with a brokerage firm in Spokane, but much to Eleanor's regret, he declined and went back to the farm. He said the people who worked in that office would catch a cold in October and not get over it until April. Heart attacks were common place. He thought they were the sickest bunch of people he ever saw. It looked unhealthy to him and he didn't want to be confined inside a building for the rest of his life.

Eleanor had never wanted to be a farmer's wife. She wanted to live in the city. One consolation was Larry's mother had one of the first Bendix washing machines and Eleanor had the use of it. They planted a ten-acre patch of oats that year and everyone watched closely all summer because the proceeds were to go for a new dish washer. So even though she was a farmer's wife, she always had an automatic clothes and dishwasher.

Without chemical weed control at that time, they worked the ground every ten days with a rod weeder to control the weeds. The soil become dry and dusty with the constant cultivation. Eleanor felt bad when Larry came in at the end of the day. After working on an open air tractor all day, he was covered with dirt so thick you couldn't tell what color he was. Eventually, Larry and his brother Bob decided to move their operations from the dryland, because it wasn't doing that well financially, and move to the Basin.

It was 1961 when Larry and a friend, Ray Emtman, bought ground in the Quincy area. Bill Hess helped them find farm ground. Hess developed DODCO (Dwarf Orchard Development Co.). He planted the first nine-acre block of orchard on Larry's unit in 1963.

Ray and Larry stayed in Quincy during the week working on the farm. They drove home on the weekends. Eleanor was a little apprehensive about the place they were staying. They called it the Doll House. Come to find out, the Doll House was on Ray's ground and the name was short for Dwarf Orchard Landlord's.

When the Richardson brothers sold the dryland farm a few years later, they picked up another three farm units near Quincy. To help ends meet, Larry did custom work for neighbors. He worked all day and part of the night cutting beans, swathing, baling and anything else he could do to buy gas and pay expenses. The units were all located west of Quincy in heavy soil that didn't drift when the winds blew like the area south of Quincy. The ground had already been developed and farmed. When originally developed, there had been no sand dunes to knock down, only minimal levelling had been required to have it ready for rill irrigation. Gravity flow took the water through the fields in corrugates spaced evenly across the fields. The water was transferred into the corrugates from a head ditch with syphon tubes.

When Eleanor moved to Quincy they bought a home in town. Larry wouldn't have to stay at the Doll House any longer. Eleanor finally made it to the city, of Quincy anyway. A lady across the street from them, brought over some cookies to welcome her and get acquainted. The lady was the wife of a local druggist. Eleanor said, "That's interesting, I worked in a drug store for awhile, if you ever need any extra help, let me know."

She was called to work the next morning. A few years later Eleanor told her friend Penny Fullerton, "If you ever leave your job at the Chamber of Commerce, I want it because I think that's the most fun job in the whole town." One day in 1965, Penny said she was leaving. Eleanor got the position. It only paid $200 a month, but she loved it. Through that position, she became acquainted with nearly everyone in the Quincy area.

Over the years they continued planting more orchard and rented other farms nearby. They finally rented ground from a large landholder, Zaser and Longston from Seattle. It was all located southeast of Quincy in the Black Sands area. They developed hundreds of acres on a development lease, getting free rent for the cost of developing the ground. That soil would blow very quickly, but it grew great potatoes and alfalfa.

They started raising potatoes in 1966. Richardson and Rudy Stetner bought a fresh packing shed from Boynton Dodge to enhance their potato sales opportunities. They only paid $500 down and so much per sack that was run through the plant. They had to rebuild the equipment lines to make it efficient. Besides running their own, they ran and sold other farmers potatoes for a set price per ton.

One of the benefits of owning Blue Ribbon was Larry and Rudy's invitation to participate in a Washington State People To People trip to China shortly after that country was opened up to visitors. The group was not only allowed to ride in the first air conditioned busses, stay in the first tourist hotels, although not completed, but to visit the many great sights of China. They also visited the communes where people had to live and work, and unfortunately, give the government a far larger share of their produce than what they received to live. It was an eye opening adventure and made them thankful to be Americans.

Larry and Eleanor had four children, Laura, Greg, Larry Jr. and Douglas. Their family has been the most important part of their lives. They were blessed with children that were artistic and especially talented in music. Things don't always turn out the way they are planned. It was a very sad period in their lives when in May of 1972, a woman ran a red light, her car hit Larry Jr. while on his motorcycle. His death came the same week he was to graduate from Gonzaga University. The school presented his diploma to Eleanor and his wife Teresa (Omlin) at the grave site. Larry III was born seven months later. Such tragedies are never to be forgotten.

Larry sold his half of the fresh shed to Stetner in 1986 and concentrated on their farming operation. His plans were to retire, but 14 years later maintains he hasn't quite accomplished it yet. He said he operated tractors many years without protection from the cold, wind and heat. Now the crew won't allow him to operate the big new tractors with cabs that have air conditioning, heaters, dust filters, radios and CD stereos. Eleanor says he wouldn't know how to run them anyway. Larry says the help should pay him for the privilege of operating those tractors in the lap of luxury compared to the older equipment.

Larry and Eleanor are very faithful in their spiritual life. They believe deeply in their religion. Not only that, they live it each day. They have made many pilgrimages to cathedrals and shrines. Their trips have covered most of the world. They are probably the most travelled people in the Quincy area.

During one period of time, they were members of a club, "the Jet Set." The club operated a 720 airplane and for a $100 apiece, could fly to Mexico. They enjoyed many trips to different locations. About 1974, on one longer than normal excursion, the plane had to refuel at Mazatlan on the return leg. At that time the airport was quite primitive. There was just a narrow strip and a small building for a terminal. A big meal was all ready for them at the Plaza, the only hotel on the beach. After eating, they returned to the plane only to find the authorities would not allow them to leave. It seems that when this plane load of wealthy tourists landed the local price of aviation fuel suddenly raised dramatically. The fuel was already in the aircraft and cash only was accepted. The captain fetched his money bag to find it was short of the demanded amount. To be able to leave, a hasty collection was made among the passengers. Among them were a few who just happened to have a lot of cash along and enough was gathered to allow them to depart. Despite that early experience, each winter they take a vacation on the western shore of Mexico for a month.

Eleanor's friend, Jane Romano, was on the Big Bend Community College Board. She was to represent the college on a trip to Europe to inspect some schools at military bases with which the college was involved. Jane tried in vain to get some of the business people, including Eleanor, to travel with her. A short time later, Eleanor got unhappy with Larry about something. One day, she picked up the phone and told Jane, "I'll go." They were gone a month to England, Germany, Spain and France.

They are very thoughtful of other people and will go the extra mile when they see a need. During the 1980's, they were part of a group of potato farmers that went to Guatemala. One farmer, Bob Holloway, from Quincy breathed in an amoeba and became very sick. They put him in the Guatemala City Hospital for several days where they were successful in treating the condition which could have been fatal. On the second day in the hospital, the time arrived for the tour group to leave. Larry and Eleanor refused to go back with the group. They wouldn't leave Agnes, the wife of the hospitalized neighbor alone in a strange land. They waited until Bob could travel and they went back together.

They have been active in community groups in Quincy through the years and Larry has served with Quincy Chemical Research for more years than he cares to remember. The task of the group was originally to do something else with potatoes beside eat them. This would be done through research and development of new technoledgy. He has made countless trips around the U.S. and even to South Africa in pursuing a beneficial climax to their effort.

Larry doesn't like to dwell on things in the past so much as what is happening now and in the future. Both of them thought the most memorable happening since the move to the Basin was their 50th wedding anniversary in 1994. It was a gala event held at Paddy McGrew's Restaurant in Quincy. It was hosted and entertained by their children with special music and song. Eleanor said it was their first wedding cake. Many relatives and friends attended the gathering including the couple who first brought them together and

Eleanor's friend from the WAVEs who stood up for them at their wedding, She came all the way from Rhode Island. They all celebrated together, the lives of a wonderful couple who had been kind enough to share their lives with so many people for those 50 years.

Eleanor and Larry

Eleanor and Larry

Larry, Laura, Larry Jr., Neal Richardson and hired hand during grain harvest of their dryland farm.

Setting syphon tubes

Mabel Thompson

A.K.A. Mrs. Democrat

It had been a long shift and both women were tired. Mabel was the owner and Myrtle, the manager, of the laundry in Ephrata. They were finishing a large order of cleaning and pressing from the nearby military base. All at once, a tremendous explosion ripped the room apart. Both women were thrown 30 feet across the room and up against the wall. An open 10 gallon container of Clorox was hurled through the air. It splattered drenching Mabel's legs and feet. They were trapped in the room with the wreckage scattered everywhere. The force of the explosion blew a large water boiler, still filled with water, through the wall, across the street and beyond the railroad tracks 300 feet away. The plumbing, which had been imbedded in cement, was still clinging to it. The two women were almost in the path of the departing boiler, but were spared a direct hit. As they shakily looked around at the surrounding devastation, they smelled smoke, a fire had started.

Dazed, Mabel got to her feet. At first, she couldn't find Myrtle, but finally saw her under a pile of rubble and helped her to her feet. They tried the outside door, but it was jammed and wouldn't open. They found a window opening and climbed out of the demolished and now burning building. They had cuts, burns, bruises and were covered with dirt and black coal dust mixed with blood which made them look like refugees from a guerrilla war zone.

It was late November in 1945. Mabel and Ben Thompson owned Ephrata's only laundry and were one of their own best customers since they also operated the Bell Hotel. The temperature was below freezing as the two women walked several blocks to the hotel where Mabel's husband, Ben, was working. Loose skin was hanging from Mabel's legs and feet where the Clorox had splashed and burned her severely. Ben wanted to take them to the hospital, but Mabel told him to go help with the fire, "We'll get ourselves to the hospital." A hotel guest took them in his auto.

Mabel remained in the hospital for several days for treatment from burns, cuts and damage done to her legs by the Clorox. Her daughter, Bertha, came to see her. When she saw her mother, she wouldn't come in the room. The four year old girl said, "Mother knows better than to have her face that dirty." Ben came in, looked at her and said, "Whatever happens, I'll still love you." Mabel didn't know what they were talking about until she looked in the mirror. Dr. Kerns had a lot of experience during the war with burns and was able to treat them in such a way to prevent scars. There was never an explanation for the blast and fire.

Mabel was born 36 years earlier to Frank and Bertha Bell in Wenatchee, Washington, on October 16, 1909. Her mother was in Wenatchee because there wasn't a doctor in the Mae Valley-Hiawatha area where they lived on a homestead. The delivery was so fast, that Mabel arrived before the doctor did. Grandmother Hill took charge, doing the work of a midwife, and handled the situation as she had many times before, in Mae Valley, for the lack of a local doctor. Four years later, Mabel became a sister. Her brother Frank was born in 1913. That made two Franks in the family, Sr. and Jr.

In 1914, Frank Sr.'s brother, Arthur and his family, were travelling from Colorado to Oklahoma for Christmas. He was in a covered wagon with his wife and two children. A kerosene stove was accidentally filled with gasoline one evening. When the stove was lit, it exploded. Everyone was thrown out of the wagon from the blast. The baby and mother were killed. Frank's brother died as a result of the burns three weeks later. That left the two-year-old boy with burns on his face and arm, as the only survivor. The boy was named Frank after Mabel's father. A younger brother of Frank's took the boy for three years before asking Frank, Sr. if he would care for him. After consulting with his wife Bertha, Frank said they would, and young Frank joined the family. Out of five, there were three Franks. To ease the confusion, Mabel's brother was nick-named, Vic.

Mabel's first year of school was at Hiawatha Valley. The teacher was scheduled to stay at homes of the valley residents for three months at a time. She liked the Bell home so well, she didn't want to move, so Mabel had her own teacher in residence. They lived a mile and a quarter from school and they walked or rode horseback. Sometimes during the winter, with snow knee deep.

The original town of Moses Lake was east of the lake on the way to Spokane. The town of Neppel was next to the lake. When Neppel grew large enough to encompass the small town of Moses Lake, the name was changed from Neppel to Moses Lake. On the west side of the lake in Mae Valley, Mabel's Uncle George had a country store. He was stocked with a little of everything from dress goods to food. Mabel compared it to a condensed, early day Walmart. Her Uncle George caught the flu from a salesman during a flu epidemic, both he and his sister died from it within a few days of each other.

Frank Bell always owned a farm, but never wanted to actually farm it. One day in 1911, he was plowing with a team of horses when he decided, "I can do better than this." He stopped the horses, went to the house and changed clothes. He walked 19 miles to Ephrata and caught the train to Seattle. In Seattle, he traded the farm for a hotel. The man trading the hotel thought Bell couldn't make a go of it and he would get the hotel back. Before it was all said and done, Frank was the one that not only made the hotel successful, but got the farm back too.

Meanwhile, back on the farm, Mabel's mom, Bertha, was left with the small dairy herd of 15 to 20 cows to milk by herself until Frank got back from Seattle. Milking that many cows by hand was a huge job, but she did it twice a day. When Frank returned, they packed up and moved to Seattle where Mabel attended Summit school in the second grade. By the third grade she was in school in Billings, Montana. Her father was there working in real-estate.

When she was in the second grade they lived in an apartment house in Seattle. Vic was about four years old and a neighborhood bully would give him a bad time whenever Vic went outside the apartment. Mabel had hit the bully a few times without any effect. One day the boy got Vic down and rubbed his face on the sidewalk. When Mabel's dad come home and heard about it he said, "The next time that happens and you don't do anything about it, I'll paddle you." A short time later, Mabel found the bully with Vic down on the ground again. This time she knew what to do, she grabbed the kid by the back of the neck and the seat of his pants and rubbed his nose on the sidewalk and roughed him up rather severely. There wasn't any problem after that.

The family was back in Grant County when Frank injured his arm severely while working under his automobile. Gasoline spilled down his arm and soon afterward someone lit a match. His arm caught fire and was badly burned. It was hard to do physical work because of the arm, so he ran for Grant County Treasurer and won. He held office from 1918 until 1923. During that time a state bank board asked Frank to transfer the county funds of $55,000, from the First National Bank to the new Grant County Bank.

The new Grant Bank went broke and Grant County lost the money. Although Frank was not responsible for the loss, he thought he might run for a political office some time in the future and decided to pay off the debt himself. It took a great deal of sacrifice, but he made up the loss over a long period of time. The County Treasurer position paid $125.00 a month so the family was short of cash for many years. Mabel remembers her Christmas present one of those years was a pair of stockings. The original bank building is still on the downtown corner of Ephrata. It has been a tavern and a pizza parlor among other things through the years.

Frank Bell was a Democrat, he raised his daughter, Mabel, to be the same. A man by the name of Dill from Spokane was running for the United States Senate on the Democratic ticket in 1923. It was the same year Frank ended his Grant County Treasurer duties. Frank had never met Dill, who had been in the House of Representatives for two terms. But Frank was active in the political process and spent time putting up "Dill for Senate" posters. One day as Frank was erecting a sign, Dill happened to drive past, stopped and got acquainted. They found a lot in common in political terms and eventually discovered they had the same birth date.

Frank ended up campaigning statewide for Dill. When he won the election, Dill hired Frank as his private secretary. Senator Dill thought North Central was the best school in the state and thought that was where Mabel should attend her first year of high school. They moved to Spokane where the three children, Mabel, Vic and Frank stayed with their Grandmother Hill the first year while their parents went to Washington, DC.

The following two years, Mabel went to high school in Washington, DC. Again, Senator Dill suggested the schools in Washington State were better and they moved back to Seattle where she graduated from Franklin High School. The Senator didn't have a family of his own and took special interest in the Bell children. Mabel did well all through high school and took extra classes each year and graduated early in December, 1926. She couldn't get her high school diploma until the rest of her class graduated in the spring of 1927. By then she was starting the third quarter at the University of Washington.

They lived on Beacon Hill where she walked a short distance each morning at 6:30 and caught a bus to the campus. One morning on the way to the bus, she saw smoke coming from a house near the bus stop. Mabel rushed to the door where two older people occupied the lower floor. They had difficulty hearing when Mabel first tried to warn them. She was told a family lived in the second story. Mabel rushed up the stairs and told the woman to grab the kids and get out because the house was burning. The woman got so excited the only thing she took with her was a picture of her children and Mabel had to evacuate the two children. When everyone was out of the house she called her mother to contact the fire department.

While attending a dance at a friend's home one evening in 1929, Mabel met a young man named Ben Thompson. He asked her to dance with him. He gave her a ride home afterward even though she stepped on his feet considerable during the evening. He forgot to get her phone number and returned to ask her to attend a movie with him. The family checked into his background before allowing Mabel to go. The movie was an early Al Jolson film at the 5th Avenue Theatre in downtown Seattle.

Mabel had a well rounded education. She majored in history and minored in English and Political Science. Through the years she attended 16 different schools before attending college and hadn't missed one day of classes the whole time. Her dad always moved over the weekend so she was on time for school the following Monday. After going to classes full time including summer quarters, she graduated from the University in late August 1929. By September 7, she started her first teaching job in Oroville, Washington.

Ben and Mabel were married on December 27, 1930. They kept the marriage a secret except for the immediate family. A week after the wedding she returned to her teaching in Oroville and didn't see Ben for five months. Ben was a department head in a Seattle Parking company making $150.00 a month which was an excellent salary. Mabel needed three years of teaching to receive a Life Teaching Certificate which she promised her mother she would do. Mabel was home during the summers except for going to summer school for her Master's Degree and the Life Certificate requirements. She taught at Oroville for three years.

Meanwhile, her father was appointed U. S. Fish Commissioner in 1933. He received some ribbing because, what would someone from a desert know about fish? Frank asked Ben to go to the Pribolif Islands in the Bering Sea to become the storekeeper for the natives there because the people in charge were not treating them right. The Pribolifs are 600 miles north of Onalaska. Ben accepted the job, and although Mabel was pregnant, he left for the Pribolifs. The doctor told Mabel not to go because of the pregnancy. They wouldn't see each other for another nine months.

Mabel had complications and eventually lost the baby in December of 1933. Ben was notified of the baby's death and that Mabel wasn't expected to live. But there wasn't anyway to get off the island.

The report was partly right, but Mabel's prognosis was a bit premature. In May of 1934 Mabel arrived at St George, one of the four Pribolif Islands, where Ben was the storekeeper. The other Islands were St Paul, Otter and Walrus. The natives harvested the seal for fur and in return, were supported by the government with food, medical care, homes, wages and schooling for the children. The beaches were crowded with big bull seals. A company from Missouri had the government contract for the dried seal skins. The program was under the Department of Interior.

The school teacher on the St. Paul Island became sick and there was no one to replace her. Mabel was qualified and took over the teaching duties and they moved there from St. George Island. Schooling wasn't a great success there, many students were 16 years old by the time they got out of the second grade. A man teacher was also needed and Mabel knew the superintendent who had been at the Oroville school was out of a job and recommended him. He was hired and Mabel felt she was able to help the one who helped her when she started teaching.

The agent on St. Paul had a heart attack and left. Ben took his place. Ben and Mabel were there for 15 months before leaving the Islands. Ben told her, " Mabel, I'm as anxious to go home as you are, but I want you to realize that our life will never be as close together again. We'll have to do something for other people instead of just for ourselves." They had bonded very close while on the islands with the lack of outside influence. Mabel said, "Being stuck on the Island, you had to kill him or love him."

Mabel's father had acquired another farm by 1934, along with 225 head of cattle. It was located in Neppel and called the Bellview Dairy. Ben and Mabel took over its management when they returned from the Pribolif Islands. They delivered twelve 10-gallon cans of milk a day to the MWAK Company at Grand Coulee Dam. Ben took care of the barns and Mabel ran the pasteurizer.

The dairy was sold. Another farm in the project was bought west of Ephrata on Martin Road. Ben spent part of his time irrigating the farm, raising alfalfa as the main crop. Mabel applied for and got the job of principal at Grand Coulee to finish six months of the school year to replace an administrator who died. She left home at 6 a.m. on Monday mornings to get to school on time and left the Grand Coulee school at 4 p.m. Friday to return to Neppel for the weekend.

In 1938, Frank built the 40 room Bell Hotel in Ephrata. Ben and Mabel agreed to run it for him. The city of Ephrata was to have a sewage system completed by the time the hotel was finished. The hotel opened in February of 1938 but the system wasn't in place until October of 1939. To handle the waste during those 18 months, three cesspools were dug 60 feet deep behind the hotel. They were amazed that digging that deep, there wasn't a single rock. Each morning, Ben filled a tanker truck with sewage for disposal. The number of loads each day depended upon the amount of guests residing there at the time.

Ben and Mabel were at the doctor's office in Seattle February of 1941. Mabel was pregnant and because of losing her first baby they ran extra tests. The doctor told her things should be fine, if you don't get any big jolts of any kind. On the way home a tire blew out and they wrecked the car. Ben's elbow was shattered and bleeding badly after the car hit a large rock. Mabel was thrown out of the car, but only had some minor cuts and bruises. The only thing available to wrap the arm was Mabel's blouse while getting Ben to the Soap Lake Hospital. The elbow needed more attention than Soap Lake was qualified to do. The railroad was contacted in Spokane and arrangements were made for the train to stop in Ephrata to pick up Ben and Mabel and transport them to Seattle. The doctor in Seattle was going to cut off his arm. Mabel said, "He can still wiggle his fingers, you're not going to cut his arm off." He was in Seattle for 10 days and came home with both arms. No harm was done to the baby.

When the hotel was built, the population of Ephrata was 800 people, plus horses and dogs. Mabel remembers a weekend when only one room was rented. But when the dam construction started, the walls were bulging with workers and company representatives. Mabel started each day washing sheets at 4 a.m. They hung curtains across the halls to divide sleeping spaces during peak periods. The telephone company personnel used the lounge, sleeping in three shifts with a change of sheets in between.

The government built the Ephrata airport in 1941 for the training of military pilots for World War II. Living quarters were built for the workers. In November of that year, Mabel received a call from a government official asking her to, "Make room for 1,200 soldiers who will arrive tonight." Mabel told him she couldn't do that. The man on the other end said, "Then we'll take the hotel over." Mabel's response was, "Young man, come ahead, if you can get 1,200 people in a 40-room hotel, you better take it over because you're a better man than I am." He became more realistic and asked her to make arrangements for the men who would have their own bedding.

Mabel went to work finding places with bathroom facilities for them to stay. She lined up the Ephrata City Hall and the basements of all the churches. When they came marching in, she had places for them to put their bedding down. That was the beginning of constant overflowing crowds of people. They saved half of the hotel for the military and the other half for salesmen connected with the dam. At times, Ben and Mabel slept in chairs all night and let others have their room. Before they started the airport construction, Ephrata had one 1,000 foot runway. After completion, there were three paved runways, the longest being 6,700 feet.

Men shared rooms with twin beds in them. One day they received a wire saying, "I'll be in on the midnight train, save me a room." Signed by A. Lyon. They made arrangements with one man to share his room with the new arrival. The only problem, that upon arrival, it was a woman. She ended up spending the night in the lobby trying to sleep in a chair.

The dam drew interest from other countries. One group came from Yugoslavia. The leader did all the talking and collected all the mail. He read all the mail first, Mabel didn't approve and told the man, "In America, we give the mail only to the addressee." They gave the mail only to those to whom it was addressed. She wasn't too happy with their presence and was glad when they moved on.

Ben and Mabel brought Sarah, a very bright 13-year-old girl, from Alaska. They raised her and put her through high school in Neppel and Ephrata. Sarah then attended the University of Washington Nursing program and later worked at Deaconess Hospital in Wenatchee.

She eventually became Supervisor of Nurses at the hospital, but died from a heart attack in 1975. Their own daughter, Bertha, was born in 1941.

The family stood on the bridge and watched as the first water come down the canal above Ephrata. It was an emotional time for them because Mabel's father had spent most of his life working to accomplish the watering of the surrounding desert. It was the fulfillment of many years of planning and work.

They started the construction of Grand Coulee Dam in 1933. Years earlier, two Democrats and two Republicans, Gail Matthews, Nat Washington, Sr., Billy Clapp and Frank Bell put up $500.00 apiece. They hired the specialist who laid out the Panama Canal to study the feasibility of Grand Coulee Dam. When working for Senator Dill, Frank was told, "This irrigation project is your baby, you go to President Roosevelt with the plan and see what you can do." Frank did a good job of convincing the president of its tremendous impact on the whole Northwestern United States, because from that presentation, the low dam was authorized. The high dam was approved later.

Ben passed away on December 7, 1956, just before his 58th birthday. Mabel ran the hotel for five more years after his death. Her brother took over running the hotel and Mabel started teaching English and drama classes at Ephrata High School in 1962. She remained there for 15 years until she was 68 years old, retiring only because of age limitations. Mabel wasn't ready to quit yet and took the job of Director of the Ephrata Senior Center for another 15 years, leaving it with good social programs and in a solid financial condition in January of 1991.

Mabel's daughter, Bertha, moved back from Seattle where she had taught at Seattle University before she left teaching and worked as a Certified Public Accountant. She opened her own CPA office in Ephrata in 1980 and Mabel has worked as her secretary and receptionist since 1992.

Mabel has continued through the years being faithful to her political upbringing. She is the Third District Committee Woman at the age of 91. In Grant County, she has been President, Vice President, Chairman, Treasurer and Secretary of the Democratic Party. She has been Washington State Committee Woman and National Committee Woman. She was a delegate to the 1996 Presidential Convention where Bill Clinton was renominated for president of the United States. She received a Lifetime Achievement Award from the Washington State Democratic Party. In Central Washington, she is known as "Mrs. Democrat." She, as with everything else she has done, wears the mantle with dignity, grace and hard work.

Bertha
Frank, Sr. Mabel
Frank, Jr. Thompson

Ben and Mabel on their wedding day.
1930

Mabel with Senator
Jackson from
Washington State

Everett and Dee Thornton

They Went That Away

It's amazing how different, and yet the same, the people were who came to the Columbia Basin in the first years of irrigation. Every person comes to a crossroad from time to time and he has to choose which direction to go.

Warren G. Harding was president and prohibition was in full swing. Model Ts were in their heyday, selling at $300 apiece. William S. Hart was a wild west hero on the silent screen along with the It Girl, Clara Bow. That was when Everett Thornton was born August 8, 1922, at home on the farm near Edgar, Montana. His parents had left Colorado from an irrigated farm to try dryland farming. His father had said that farming irrigated land was crazy, he was tired of it. Everett was one of eight children, six boys and two girls.

Everett had two years of college in York, Nebraska. School was interrupted because of the attack on Pearl Harbor in Hawaii on December 7, 1941. He enlisted in the U.S. Army Air Force in 1942. By the time he was called to active duty it was February 1943.

After basic and flight training he was assigned to the 303rd Bomb group at Molesworth, England. It was one of many 8th Air Force bases clustered in an area about 75 miles north of London. This was the same group that the famous actor, Jimmy Stewart, had served in during the earlier stages of the war.

Everett flew 23 missions across the channel as either pilot or co-pilot. During that period, January to April of 1945, German fighter planes were virtually nonexistent and anti-aircraft fire did little damage at 30,000 feet. Nevertheless, the weather did provide a challenge. England is known for its foggy weather and Everett had the misfortune to witness it first hand on more occasions than he liked. They took off on part of their missions by setting the gyro compass and headed down the runway for takeoff. Flying in close formation in the clouds was also tricky. It took both pilots constantly monitoring to keep their wing tips from hitting the B-17 bombers next to them.

His discharge from the service came in September of 1945. The first place he headed for was Montana and the farm. It took several years, but he finally finished his schooling after attending different colleges little at a time. The graduation was from Denver University.

By the time Dee Bergloff was born December 8, 1928 in Bismark, North Dakota, Calvin Coolidge was President. That was the same year that Micky Mouse appeared for the first time as the hero in "Steamboat Willie" and Henry Ford brought forth the Model A. Among the celebrities that year were Tom Mix, Gary Cooper, Charlie Chaplin, Al Jolson and Al Capone. Silent pictures were on the way out as the "talkies" were coming on strong.

Dee's father was an electrician. He had a job with Northern Pacific Railroad putting in power lines. He traveled the Northwest States doing construction for the Northern Pacific before settling in Billings to raise his family. At the tender age of six months, Dee was moved to Billings, Montana. She was raised in Billings and met Everett in the United Methodist Church where their families both attended.

After high school she had one year of college at LeMars, Iowa, and three years of nurses training in Bismark where she graduated in 1951. The college atmosphere wasn't as noisy in those days, but still bad enough that she hid out in the furnace room the first year to concentrate on her studies.

After graduation she returned to Billings where she worked as an Assistant Nursing Arts Instructor, following student nurses around to make certain they were doing what they should be doing on their floor assignments. The last year and a half she spent working as a supervisor-instructor in pediatrics. Nursing must not have been enough of a challenge for her because in May of 1953, she married Everett.

Everett wasn't the only one of his brothers and sisters to get married. It became obvious the farm could not support them all. Through a newspaper ad, Everett had contact with a sawmill owner in Everett, Washington, who was looking for someone to invest in his business. It was a good chance for a winter trip and the newly married couple packed and headed west. On the way to the coast, they passed through the Columbia Basin. They had heard the publicity about the irrigation project and the veterans drawing for free land.

When they arrived at Everett, Washington, they were not satisfied with the sawmill owner's plans that were laid out for them to invest in the operation. They declined the offer and decided to go south through Oregon and ended up in California. Everett got a job with a building contractor, but pounding nails day after day didn't give him much satisfaction. Two years later, on the first of February 1955, with their daughter Linda and all their worldly possessions in a small home made trailer, they headed north.

They decided Everett would return to school, this time at Washington State College at Pullman. He was considering working for the Extension Service as a career. After driving all night they were tired. They stopped in the early hours of the morning at a junction south of Pasco where they made the most important decision of their lives. It was there on that cold winter morning they discussed the pros and cons. Should they go to Washington State College or take another look at Quincy and the new irrigation project? They were at a crossroads, not only for direction the road went, but where their future would be. They decided to abandon the college registration that coming Friday and headed for the desert with a promise of water.

Housing was not available in Quincy. While they were looking and talking about locating farm ground, they found temporary housing at the Garland Motel in Soap Lake for the first two months. By May they found a house in Quincy across the street from the junior high school.

Meanwhile they found a quarter section of land that had been partially farmed and owned by a realtor from Ephrata. They paid $27,000 for 160 acres and a small amount of old equipment, a few hand line sprinklers and a pump.

Everett said, "He was probably sure he would get it back, because he didn't know how much money I could stand to lose."

They were able to start planting hay in the spring of 1955. With the help of a good neighbor, Jack English, who had worked with the previous owner, and plenty of advice from other neighbors, they gradually developed the farm over several seasons. The 160 acres was on Road 8 NW, a half mile east of the Adams Road.

A hand line sprinkler was the method of irrigating at first. Everett was in good physical shape. He could change a quarter mile of three inch lines in 27 minutes. It's good he was fast because he did it three times a day and there were probably at least four lines on the 160 acres.

After several years he changed to wheel lines. They were fortunate to eventually buy two 80-acre units between his original ground and the Adams Road. It gave them 320 acres in a block. With two quarter sections, it was natural to install two circle systems when they became available. The circles were among the first installed in the area. It has been a great place to raise their four children, Linda, Jack, Mark and Jane. Mark has taken over the farming.

They built a house in 1956. Building costs were much cheaper in those days. They moved into a solid, well-insulated house at a cost of $9,000. A generous neighbor had prepared the ground for a foundation. Mike Hale and Gene Shulz provided good workmanship. Quincy Lumber was the source of supply for the materials. The siding wasn't on yet and some trim and mouldings weren't in place, but it got them on their own property. Irvin Cizik finished the siding for them several years later as finances permitted.

They didn't have any grass around the house for several years. They wanted to wait until they could install underground sprinklers.

When they built the new Presbyterian church in 1959, they planted grass around the new building. Grass seed was left over. The minister, Larry Roumpf, appeared at the farm one day and said, "You've been without lawn long enough, lets get this grass in."

Alfalfa was the main crop through the years. Along with the farming, Everett used to load his own hay and haul it to market west of the mountains. He did raise a few other crops including beans and for three or four years, potatoes. The first year in potatoes, they made very good money, the following years took it all back.

Dee remembers the long rows of beans that she and the children hand weeded, and the weed throwing fights that occurred when their own and the neighbor's kids got bored.

One of those years, the government had a "Potato Diversion" program. Prices were so low many farmers would have gone bankrupt. To qualify for the payment, the potatoes were to be destroyed or go for cattle feed. To insure the potatoes would not find their way back to the marketplace, they nicked them mechanically. It was winter time, the temperature cold. A converted barrel hay chopper was used to nick the potatoes and the equipment was set up on the corner of Everett's field. Several local farmers were running their spuds through the machine and hauled to a nearby feed lot on Adams Road. Seven or eight farmers, all bundled up to keep warm, were standing in a circle around an open fire visiting. Among them was the government USDA inspector, Bud Henson, from Quincy. They were all talking when they heard Bud yell. All eyes went to him as he jumped back from where Everett's dog was standing with his leg raised. It had just completed a thorough wetting of Bud's pant leg and boot.

"Who's dog is that? Bud indignantly demanded.

"I don't know." Everett said.

No one could bear any other comments because they were doubled up with laughter for the next five minutes. Later Everett said, "The laugh was worth the loss on that bunch of spuds." A few of those years, fun was the only profit farmers had.

In her spare time from the farm, Dee worked at her profession in both the Soap Lake and Ephrata Hospitals. She did off duty nursing at times and taught Health Occupation for three years at Quincy High School for Big Bend Community College.

The Thorntons have attended and been active in the First Presbyterian Church since coming to Quincy. They especially enjoyed a group within the church called The Mariners. Dee recalls the good times spent with other young couples in their activities. Pastor John Christiansen and his wife Fran formed the group in the sixties. They eventually disbanded it as the young couples became busier with families and the wives having to take jobs.

They have been active in two other groups for many years, called Veterans for Peace and Pastors for Peace. It was 1997 when Everett volunteered to drive a donated school bus to South America for them. These groups support a university system with campuses in three small communities in Nicaragua. Everett flew to Chicago in November to start his journey south with a member of Ojibwa Nation, Ann Duirr, from Minnesota as navigator.

The first destination was in the far corner of New York State to pick up a special hospital bed. Upon arriving they found the bed was too large to get in the door of the bus and had to leave without it. It wasn't long before they were completely loaded at stops along the way. Their cargo was medical supplies, hospital beds, books and computers. The cargo was held back from the driving portion of the bus by restraining ropes which would not have held if they had a sudden stop.

Four thousand miles, eleven days and ten meetings later they arrived in San Antonio, Texas. Ann returned to her home. She had completed her part of the job. Everett's part of the trip had a long way to go. Joining him were two other buses, two trucks, two ambulances, two pickups and 21 other very wonderful people. The large hospital bed they couldn't get in the bus in New York State was there, arriving by freight line. It was loaded on one of the trucks. Most of the personnel were from the U.S. However, there were two Canadian women and one doctor from Nova Scotia who were going to work in the San Christobal Hospital for two weeks. Others of the group were to stay for a few weeks doing hospital maintenance, improving water systems and planting trees.

The caravan went south down the east coast of Mexico. The organizations refused to pay any bribes. Two Mexican Immigration officials, probably with expenses paid, traveled through the country with them, rushing their way past military check points. They drove through El Salvador, Honduras, Guatemala and finally to Managua, Nicaragua.

Everett had looked forward to helping distribute the supplies, but upon arrival, they were required to deliver everything for impoundment. He understood they released it ten days later. Everett was in Managua and along the East Coast several more days, visiting some poverty stricken communities before flying from Managua for home.

Meanwhile back to the farm, Everett maintains he was a terrible manager and not cut out to be a farmer, that he was meant to be a truck driver. Somehow he managed to do both. Everett has always been willing to try something new. That, along with hard work and determination was the difference between success and failure. Many others failed who didn't work as hard and couldn't learn something new.

When he was 57 years old, he tried surf boarding on the ocean in Hawaii. It was at least ten years later when he tried flying a powered parachute at an airport near McMinville, Oregon. Everett said the sales agent wanted to make a sale so bad, "he had me over dressed and under schooled and under prepared." Everett was fitted with a helmet and earphones for verbal instructions during take off and landings. He was strapped into the open seat of the three-wheeled frame with a propeller driven engine mounted two feet behind him. With no other directions, he was put on the runway and told to take off.

"And I was dumb enough to do it," Everett said.

The chute came up according to plan, not knowing how the steering worked, he inadvertently steered to one side and trying to correct his direction, made it worse. He was going off the side of the runway with buildings and power line ahead. He cut the power, but not in time to keep him from getting pulled over sideways and bruised up.

Everett said, "I was lucky I didn't get killed in that thing"

One of the most memorable days over the years was May 18, 1980. It was a Sunday morning when the sky looked different than anyone could remember. Something was wrong, the daylight started to fade away. Those who were outdoors saw ominous appearing clouds forming that turned a dark color with a billowing under surface. The birds quit singing, everything was almost silent.

Those who were watching TV or listening to a radio knew what was coming. But those who were outdoors or in church only knew it was getting dark in the middle of the day when it was supposed to be light. Everett and Dee thought the clouds were so odd, they got their camera and took the accompanying picture as the cloud approached the farmstead.

Some people thought it was exciting, some were rejoicing that the end of the world was here and Christ was coming, others were fearful for the same reason. Thirty minutes after they took the picture, it was dark. It was then they saw a very soft, light dust started to fall from the sky.

By this time nearly everyone knew that Mt. St. Helens had literally blown its stack. A huge explosion ripped the top third of the mountain into a fine dust that would fall as far away as Spokane and beyond. The ground on the Thornton farm was covered with volcanic ash at least two inches deep. At Ritzville, 60 miles north, it was six to eight inches. Travelers were stranded for days in some towns in Eastern Washington.

The ash was so fine a car driving on the road would cause it to raise in a cloud that could blind the driver of another vehicle. A neighbor, Jim Hirai, was killed when an oncoming car raised so much ash that another car following couldn't see, resulting in a fatal head on crash. The ash ran off the windshield almost like water. Trying to scoop or scrape it was difficult because it was so light and fluffy. Vehicle engines were destroyed if the ash was not properly filtered. Many crops were buried as they were still in early stages of growth. Potatoes fared well and came up through it. Most alfalfa fields were in bad shape and unsaleable. The reason being, if they separated the ash from the hay, many buyers wouldn't buy from the area for fear the ash residue might harm their animals.

Everett and Dee retired from farming, except for helping Mark occasionally. Too much time on his hands is not Everett's style. At 78 years young, he is still driving truck hauling apples from Quincy to Brewster and Wenatchee and tree seed stock to Portland. During the spring he brings in potato seed from Lynden and even hauls hay occasionally, including helping with the loading.

He says, "I've got to do something and it brings in a little to help support my favorite organizations, Pastors for Peace and Veterans for Peace.

What stands out in their minds now after all the years in the Basin? Their answer, "Being associated with such nice people, there have been outstanding, nice people in this country and that's all there is to it. It's a great country to live in." Everett said "Anyone in farming must be grateful for surviving so many years without getting seriously hurt or killed." His only regrets are that farming is not as profitable as it should be. The low returns have caused farm acreage size to go beyond the intended acreage limit of 160 as originally planned. Nevertheless, they have never regretted the decision when they went that way at the crossroads.

The Thornton Family

The sky, May 18, 1980

Plowing down the volcanic ash

Ed and Signe Williamson

Cattle In The Blood

President Grover Wilson had to deal with many problems during and after World War I. Two things happened in the year of 1918 that affected everyone in the United States. The surrender of Germany in Europe on November 11, and a flu epidemic in the US which killed half a million people including half as many troops at home as were killed in battle. However, out in Bend, Oregon, two separate events happened that went almost unnoticed elsewhere, the birth of two babies, Ed and Signe.

Ed was raised on a dairy. He said his father, Charlie, didn't teach him how to milk until he was almost five years old. Charlie had a small heifer he was going to butcher because she was too small. Ed talked him out of killing her. Charlie said, "Alright, then you milk her". His dad tied her outside and told him, "go at it". By the time the session was over, he had more manure in the bucket than milk. That heifer ended up being one of the best in the herd.

Signe and Ed went through high school, met and were married in April of 1939. They started farming near Bend. They worked with dairy cattle and raised wheat and hay until selling out in 1945. Ed helped move the dairy cattle to California for the new buyer and said that was the last of dairy cattle for him.

Their two oldest sons, Larry and Jerry, were born in Oregon. After selling the farm, they all moved to Sunnyside, Washington. It wasn't long before the ex-dairyman was back in the dairy business again. This time they worked for seven years before moving to Quincy, Washington. The two-time ex-dairyman finally quit milking cows and started a feedlot.

The new place was on the corner of what is now Adams Road and 6 NW. When they first arrived, Adams Road was just a cattle trail. Grant County paved it during the first year of their residence. The farm was ten miles southeast of town. It was 1952 when they delivered the first water to new farm units. Starting in sagebrush and jackrabbits, they constructed corrals, a farmstead and leveled the ground into fields. A new well, septic tank and drain field were installed. They put up a new brick home which was unusual for new arrivals out in the sagebrush.

Ed was about six foot two inches tall and had brown hair and eyes. For a gentle, soft mannered man, he was an extremely hard and tireless worker. For a work partner, Ed met his match with Signe. She was five foot five inches and had brown hair and eyes. She evidently had boundless energy because she raised four children while doing countless hours of field work on the tractor and by hand. One of her neighbors, Ruby Holloway said, "Signe was the only woman I knew that worked on the farm as hard and long as I did."

The fields were leveled for ditch gravity irrigation at first. The soil was sandy and keeping it from blowing out crops of wheat and alfalfa was difficult. Signe said she even used tumbleweeds to mix with the soil trying to repair breaks in the dirt head ditches that delivered water to the corrugates. At first, horses belonging to a large landowner, Paul Lauzier, were still on open range. They would occasionally walk on the soft, sandy ditch banks and cause more breaks. The sand had no humus and the dirt bank would almost melt when water soaked up the moisture.

A neighbor's wife, frustrated trying to stop a break in the ditch, just sat down in the gap until someone came along to shovel it shut. Another neighbor, with a wider break, had his teenage son lay lengthwise across the opening while he shoveled the opening closed. As the years passed, farms keeping the gravity flow irrigation, eventually put in cement ditch linings, which also helped to conserve water.

It wasn't long before they put handline sprinklers on the fields which helped control the blowing problem. The sprinklers had to be moved often, usually twice a day. Each length of pipe was 40 feet long and four inches in diameter. There was a two and a half foot, three-quarter inch pipe in the middle with a impact sprinkler head to disperse the water. The lines were connected end to end completely across the field. They were joined to a six to ten inch diameter main line every 60 feet. Each line had about 10 settings apiece, giving a coverage of the field every five days.

A few years later the wheel sprinkler lines were available. The first wheel lines had to be rolled by hand to the next setting. It was still a family effort to change water. Eventually gasoline motors were used to drive the wheel lines and could be moved by one person.

Wilson Creek was the next stop for the Williamsons. It was 1956 when they moved onto the ranch where they raised cattle, wheat and hay. They were about the first in Central Washington to use a Harobed automatic haystacker. I remember my father and I going there to see it before I got my first one.

The youngest son, Alan, decided to arrive in 1957 during the worst flood in Wilson Creek's history. Both Wilson Creek and Crab Creek flooded over their banks causing havoc in the small towns in the area. The high water cut Wilson Creek off from the highway and the hospital. The only communication was by telephone.

Ed knew he couldn't get Signe across the flooded road by car. He thought he could make it in the truck if he could manage to stay on the high roadbed. The tops of the telephone poles were sticking up only a few feet above the water along side of the road. They slowly felt their way, having to back up when Ed felt the front tire starting to go off the shoulder.

A helicopter had been alerted just in case they didn't make it through. Fifty people watched from the other side giving moral support. After all the difficulties, they finally arrived on the other side to the cheers of the onlookers. Alan had made his first trip, it's a wonder he didn't become a sailor.

Woodburn, Oregon, was the next stop after selling out at Wilson Creek in 1961. They started clearing ground again and gathering crop contracts. It was still 1961 when they made a deal for ground on what they called the Reynolds place, south of Winchester, Washington. The new farm was on the corner of roads F and 7 NW. The cattle feedlot they put together expanded to 200 head during the six-year stay at that location. While on that farm, the Ephrata Chamber of Commerce awarded the Ed Williamson family, "Farmer of the Year." The award was presented by Stu Bledsoe, our Representative to Olympia at the time.

Signe said one winter night, a band of dogs chased the cattle through a fence. It was a real mess, they had cattle scattered all over the nearby countryside. It took two days to get them all rounded up again. One neighbor by the name of Cook owned two of the dogs, which he promptly shot.

Ten years went by before they sold again. They moved to Ephrata in the year of 1971 where their daughter, Marcia, finished high school. Meanwhile, they were building a house south of Interstate 90 near George. After Marcia graduated, they moved back to the country.

Ed's father, Charlie, was a character. Everyone knew him. He was a big man, always good natured and friendly. Charley was probably the roughest talking man most people ever met. He could turn the air blue with his language. Nevertheless, he was like finding water in the desert, because that is exactly what he did. He had the vision to see what the new circle sprinkler could do in the otherwise useless sand dunes. He bought a large amount of ground south of Interstate 90 near Adams Road. In the late 1960s, he drilled wells and installed irrigation circles covering 125 acres each. Most people thought he was making a mistake trying to farm that worthless sand. He not only proved it was a good move, but set off development of what they call the Black Sands farming area.

It was in the Black Sands area, where Ed and Signe started to clear ground, drill wells and build another new red brick home. This time it was for a longer stay. It has been 28 years since they moved into what is still their present home. When the house was up, but still without windows, the wind started to blow one day. The sand was a foot deep inside the room after the winds went down and had to be shoveled back out through the windows.

Ed and Signe finally retired from active farming, but their children are all living in the area and are involved in rural life. Larry and wife Kathy, Alan and Melody, Jerry and Diana are all farming and active in agricultural organizations. The daughter Marcia, keeps all the books for the farm, while her husband, Erik, works as an architect.

Ed and Signe didn't make worldwide news when they were born. But they have done more than their share in harnessing the potential of the desert. They not only developed the land, but also the type of children to care and nurture the future of that land.

Ed and Signe Williamson and their new
John Deere

Feburary 1957 - Wilson Creek was flooding at the time
Signe was ready

Larry, Kathy, Signe and Ed accepted Farmer
of the Year award in Ephrata from
Representative Stu Bledsoe

Don and Byrdeen Worley

There's Got To Be A Better Way

The bitter cold north wind was whipping past in gusts that took his breath away. The platform was coated with ice. He calculated every step to avoid slipping from his perch hovering 70 feet above the frozen ground below. The oil rig tower man could feel the penetrating cold stiffen his fingers as he struggled to accomplish his task. It was too cold to snow on that western Nebraska day and Don Coleman Worley thought, "There's got to be a better way to make a living than this."

For a year, Don became a roughneck, working in the oil fields. There was something wrong about working high up on an oil derrick during winter storms with icing and a high, cold wind. Somehow it leaves a lot to be desired with your situation and makes one think about other pursuits in life. Nurses at the local hospital knew how oil rig accidents came in three categories. The first being the loss of a hand or fingers when a chain is thrown around a pipe and tightened. The second was a fall from a smaller rig resulting in back injuries, usually a paraplegic situation. The third was a fall from a high platform, ending the job with a ride to the mortuary. He decided using the G. I. Bill for an education might not be so bad after all.

Don was born in 1930 on a ranch west of Alliance. He was the second of three children. Their parents were Harvey and Helen Louise Worley. Don was the third generation of Worleys in Alliance, Nebraska. His grandfather ran cattle on open range from Alliance east to the Sand Hills to near Valentine, and from Platte River to South Dakota. They had to be strong men to care for the herd and keep peace with other ranchers. They had to keep the cattle from the big operations who liked to help themselves to small herds or strays. Roundups and brand sorting in the fall were a multi-ranch occasion.

Don's mother, Helen Coleman, was born in Sacramento, California. Her father moved to Seattle where Helen graduated from Queen Anne High School. The next stop was Quincy, Washington. She received her teaching certificate in Ellensburg at the Washington State Teachers Institute. Helen taught in a country school at Winchester located east of Quincy. At the beginning of World War I, she joined the Red Cross and was stationed in France. Meanwhile, her parents moved to her aunt's homestead and a sod home near Alliance. Upon returning to the United States, she updated her teaching certificate and taught in several country schools near Alliance. That was when she met and married Harvey.

Don's first eight school years were in a one room country school. His high school was at the University of Nebraska School of Agriculture in Curtis. Curtis was the first high school to hold rodeo competition. Like most of the young men around the ranching area, Don tried his hand at rodeo events. He did some calf roping and bareback bronc riding. He entered the steer-wrestling event in one rodeo, but after missing the steer and eating a chunk of Nebraska dirt, he dropped that event. Don and Byrdeen became acquainted during their school activities before Don graduated in 1948 and Byrdeen in 1949. Don's first year of school at the University of Nebraska revolved around drinking beer and playing pool. By the end of that first year, his parents thought he should discover how much work it takes to earn a dollar to play pool.

He went home and being young and energetic, he and a friend, joined the U.S. Army for two years. Don served part of his time in Korea. He thought the military service was a good learning experience, but not what he wanted as a career. He impressed his superiors with his sense of direction and spent time as a message runner. His sense of direction may have been good, but it didn't keep him from falling into a Korean honey pot once. He discovered they don't fill them with honey. He also acted as a driver for officers in the field. Don didn't feel he was in harm's way too much, other than running over a land mine with the jeep. Upon his discharge from the army, he returned to Nebraska.

When Don returned from the military, he wanted to improve their hay meadows by drilling wells for irrigation water. He bought cattle and things were going great until the hay grinder burned out a bearing one night. The resulting fire burned all of his hay. Without feed, he had to sell the cattle. At least the money from the cattle sale was enough to pay off the bank. He found himself broke and went to work in the oil fields for a year.

Byrdeen Carlson was born at her grandmother's home in 1931 at Potter, Nebraska. She was the eldest of four children and the third generation of homesteaders who arrived from Sweden and Germany in 1892. Grandfather Carlson helped build the Union Pacific rail line, worked in gold mines near Central City, Colorado and the smelters in Smelterville to pay for the homestead. Emil (Ted), Byrdeen's father, was born on the homestead. Emily Hagemeister, Byrdeen's mother, was born in Colorado. The Hagemeisters moved from Germany to the U.S. to escape the pending World War I. They didn't want their sons serving in the Kaiser's army. Emily's family moved to Potter, Nebraska, where she met Emil. The Hagemeister family owned several grain elevators in Western Nebraska where they farmed. Byrdeen's brother still operates the family farm which has been in the family for more than 100 years, since 1895.

Dr. Miller who delivered Byrdeen, accidentally shot himself in both feet while hunting, resulting in the amputation of both feet. Because of the disability he entered politics and became a United

States Senator. He was on hand during the Truman presidency when a few Cubans launched an attack in the U.S. Capitol, wounding several people. The doctor had some national press at the time because he gave emergency treatment to the wounded on the spot

Byrdeen's first eight years of education were in a rural, one room school. After attending one year at Potter, she transferred to the University, Nebraska School of Agriculture at Curtis. The school was set up for farm and ranch students who had to travel great distances to attend their local high schools. Students were room and boarded campus style for the school year. They provided separate dorms for the boys and the girls. She felt it was a maturing experience. Byrdeen had many social activities including, science club, dorm and student government. They had great supervision and everyone had responsibilities on the campus and the school farm. The school had a horse and dairy farm where the boys worked part-time gaining on-the-job training. Their school year was eight months long so the young people were out in time for spring work at home on the farm and ranch.

After graduating from high school, she attended the Sisters of St. Francis School of Nursing in Alliance, Nebraska. To complete their training, the nursing students travelled to St. Louis for pediatrics. They also spent time at Montana State psychiatric and tuberculosis hospitals. Byrdeen thought, for a small training school, those travel experiences were of great value. After graduating, she worked in Sidney several years before entering the University of Nebraska.

The following is Byrdeen's account of rural medical care in the 1940's and 1950's.

Being raised in the rural Midwest, one learns a basic first law that your word is your bond. While caring for a young accident patient, Dick, wearing a full body cast, teased me about going on a date. Finally, one day I said, "Dick, if you can get out of the hospital in this body cast, I'll go on a date with you." About three days later, I was told the promised date was on for a movie at the local drive-in. I was to meet him at a certain time the next evening. What a date!

The next evening, the local mortician and his wife pulled up in their new, shining black, hearse-ambulance at the meeting place, the side entrance of the hospital. The on duty firemen loaded Dick, body cast and all, in the back. Dick's mother was along. She was a local business woman and a delight, besides, she was paying the bill. It was not a quiet event for the small town. We did enjoy the evening. Dick got rid of some of his cabin fever from being cooped up in the hospital with his cast. The guys from the fire department gave me a bad time about dating heavy patients. The experience taught me to plan ahead and think before glibly accepting dates.

The hospital I worked in was an old bank building. The walk-in safe was our x-ray room. If we had more than two surgeries, we boiled our instruments. The syringes and needles were boiled for sterilization. All the needles were not new, nurses watched to keep them sharp. Our linens for surgery and O.B. were autoclaved by an old monster. Our surgery area was tiny and we used a lot of ether and spinal anesthesia. Later, sodium pentothal was a godsend. When our new hospital was built, there were many happy people. On nights when things were crazy, or we were very short of staff, the doctors would have the husbands give their wives a few drips of chloroform in the delivery room. This really shocked this young nurse, but I soon learned we had wonderful, sensible and knowledgeable rural patients. The physicians in Denver marveled at the lives our doctors saved and the ingenuity they had, considering the lack of equipment.

Byrdeen saw Don occasionally Friday or Saturday nights during her stay in Sidney. He would stop, visit and play Canasta with her, the landlady and the housekeeper. Don and Byrdeen called themselves the old bachelor and the old maid. She decided to return to school at the University of Nebraska about the same time Don started again. They accidently met while in line for registration and they began meeting for coffee every afternoon.

They were still attending the University in 1957 and both were a bit discouraged. Don was tired of living in the dorm with juveniles and with the high cost of living in apartments. Byrdeen's plans had been moving too slow. She thought about going to work for an oil company in Iran or other parts of the world. That must have panicked Don, because he proposed a plan to sell his combine and buy an engagement ring. Byrdeen said, "I don't know." Don had been her long time friend and confidant. So she said, "I guess it would be all right." A very romantic moment. They were married February 1, 1957. However they always celebrated their anniversary on February 2 (groundhog's day). It was easier to remember. Later, at a bridal shower, she enjoyed shocking her aunt by announcing, "I had to get married. Don had a two-headed coin and he always flipped for coffee, and won. It's just too expensive buying coffee for him all the time."

They planned to come to Washington after Don's graduation in January 1959. Don's mother had loved the country around Quincy during the short time she lived there in the past. The news of the irrigation water becoming available for the desert ground intrigued them. Perhaps Washington would offer another chance to build something for themselves. They wanted to be away from family and make it on their own. After all, between them, they were related to almost everyone in Western Nebraska. Armed with the stories from Don's mother, they had made one exploratory trip the year before. This time they gathered all their worldly possessions in their car and started westward in earnest.

They stayed with relatives and friends along the way through Nebraska. As they said their goodbyes, friends told them they would soon be back. Byrdeen's response was, "No, we won't, we only have enough money to make it there, this is a one way trip."

Their first stop in Washington in January of 1959 was at WSU at Pullman. When stepping out of the car, they found a one dollar bill in the gutter. Possibly a good omen! Don was considering a position with the extension service or work on a Master's degree in more agriculture economics. Two other possibilities were positions in real estate or banking.

Spokane was the next stop. Employment was available for nurses everywhere, but not much opportunity for Don. They were referred to Clark-Jennings & Associates. He met a company official, Kit Carson. Kit said later he was sold on hiring Don after seeing Byrdeen knitting out in the car with a suitcase and two tires on top. They were sent to Yakima and interviewed by Andre Charvet. Don became a Clark-Jennings employee as a mortgage loan professional. Meanwhile, Byrdeen started working at the Yakima General Hospital.

Don's area soon became the Columbia Basin and all of Grant County. His business vehicle was a small Metropolitan car. Upon one of his calls, a friendly St. Bernard as large as the car, unnerved him by sticking his head in through the window and onto Don's lap. He was soon driving a new Volkswagen. He drove all over Grant County in that little Volkswagen with his maps and compass to find his way around the sagebrush and country dirt roads. He ventured forth to help refinance farms in the Basin. The Moses Lake branch of Clark-Jennings opened in September of 1960.

Their first child, Scott Harvey Worley, was born November 1960. They found things were interesting at times with a new baby in the office at their home. Don was with Clark-Jennings for nine years before starting his own business in 1968. They called it "Farm Management Services." Meanwhile, Byrdeen was working at the Moses Lake Hospital. One day in 1968, she was in surgery when a doctor told her a baby was coming up for adoption. Would she be interested? She said, "Yes, I'll go home and talk to Don about it." When she informed Don a baby was available, he said, "You told him yes didn't you?"

"I thought we should talk about it." She said. "What's to talk about?" said Don.

The next day at the hospital, Doctor Young said he hadn't heard from her yet, did she want that baby or not? Byrdeen said, "Yes, we want the baby." After some verbal bantering the doctor started to leave through the exit when Byrdeen called after him and asked, "Hey, what did I have, a boy or a girl?" That's how their daughter Kristine Dianne entered their life a few days later. Byrdeen resigned from the hospital to care for the newest member of the family.

Metropolitan Mortgage in Spokane was interested in investment opportunities in the Basin. Don checked at the Bureau of Reclamation concerning their interest in property south of Interstate 90 from

George to Dodson Road. The Bureau said the ground is worthless and they had no interest in it. He could do as he pleased. The State of Washington said they would give well permits for the area. The Bureau was trying to take over the water rights from the state. Metropolitan Mortgage put up one million dollars toward the development. Don kept land levellers and well drillers working for almost two years.

After the wells were flowing, the Bureau of Reclamation claimed the shallow underground water belonged to them. Don had lined up several farmers to purchase portions of the development. Bankers told the buyers they would lend them money the following spring to purchase the ground and pay production costs. Early the next year, Don came home sick from what the bankers did to the farmers. The banks refused the promised loans. Byrdeen asked what would happen to the land. He said it would just have to lay and let the sand blow. Byrdeen replied, "We can't do that to the land. What can we do?" They were both about in tears when Don said, "What do you think about us borrowing the money and see if we can farm it?"

Byrdeen said, "Your hand shakes when you sign your first million-dollar loan. It was 1971 when they bought tractors, trucks, machinery and put up some headquarters buildings. They were way over their heads in work and debt, big time. They set up to farm thousands of acres. They managed to lease some circles to put on the ground and planted alfalfa, wheat and potatoes. The potato price was good that year. They were able to finish the new house they were building and pay off the million-dollar farm mortgage.

The Bureau of Reclamation stepped in and said they could not farm that many acres because of an acreage limitation. The Bureau said no one could have more than 1200 acres in their name. Everyone in the real-estate office and Byrdeen's brother all had land in their names. Don and Byrdeen still had all the risk, but at least they salvaged the new farm ground by getting it into a crop.

They worked hard to find buyers and sell the ground to meet the land limitations. Different parcels were sold to Stanford University professors, auto dealers in California and Seattle business people. Most took advantage of IRS rules at the time for farm tax write-offs.

The new owners needed someone to manage their interests. Worleys helped them form individual partnerships and leased the ground from them. Don had his hands full getting all the proper paperwork completed with the Bureau and setting up operating loans for the individual corporations. Darrell Reese, a local attorney, helped guide him through the maze of paperwork and Jeff Fagg from Moses Lake was, and continues to be, their banker and loan officer.

Prudential bought 3000 acres at Lind Coulee with the understanding that Farm Management Services would manage them. As a favor to Prudential, they took on 3000 more acres on the Snake River, but only for one year. Nine-thousand was too many acres and too much stress. There were times when Byrdeen was perhaps the steadying influence for Don during the trying times of his life. She was always there to encourage him when he was down and back him up if he got in over his head. Don toured every dam on the Snake and Columbia Rivers. He was totally in love with the northwest, especially the farming and the farm families.

They had good times along the way, even with the pressure that went with the farming and business enterprises. Don loved to cook. It made no difference if it were a gourmet meal or an outdoor barbecue, he was excited looking for the right recipe for the occasion. He loved picnics, office parties and holidays. He personally stuffed every Thanksgiving turkey.

In spite of all the activity of the loan, real-estate and farming business, Don was active in several organizations. Among them were The Columbia Basin Development League, American Society of Farm Managers and Rural Appraisers, Black Sands Irrigation Board, Moses Lake Port District and Big Bend Community College Board Member. He served as a Deacon in the Presbyterian Church and taught a Sunday school class for two-year-olds.

Don and Byrdeen's parents made a trip to the Basin from Nebraska together once and after looking the area over said, "We never saw such barren, godforsaken country in our lives, but the kids are happy." Before long, they ended up moving here too.

Don and Mr. Lee, from Ephrata, owned ground in Moses Lake. They offered it to the city for a police center and jail. A few days later one of Don's real-estate friends called and told him. "Don't go to Ephrata today. The word got out. That damned Don Worley is trying to take care of the jail needs now and the next thing will be to steal the courthouse for Moses Lake.

Perhaps someone in Ephrata knew of Don's origin in Nebraska many years before. Henningford was the county seat of Box Butte County. There was a running battle between Henningford and Alliance disputing the best site for the courthouse. The courthouse consisted of an older home. Late one night a group from Alliance stole the courthouse off its foundation in Henningford and moved it to the railroad tracks. It was loaded on a flatcar and transferred 18 miles south through the dark to Alliance where the county seat has been ever since. Older Henningford people were still angry about it 50 years later.

When Don had a fatal heart attack January 8th 1987, Byrdeen wasn't familiar with the makeup of the partnerships and corporations which they had formed over the preceding 16 years. While Don was building a complex business enterprise, she had been busy working in the Moses Lake Hospital and raising two children. She had a very loyal crew led by Arnold Greenwalt and Petra Hovland, her office manager. They stuck by and helped keep things together while, with the help of Darrell Reese, she learned the intricacies of the operation. Because of the experience of her crew, Carnation Company allowed her to maintain their potato contracts. At that time, they were Carnation's largest contract grower. She wanted to learn all she could about what the crew was doing. Arnie and the men were very patient in getting her acquainted with different aspects of running the farm.

One day Byrdeen was sitting in Don's office. She didn't know if she was half asleep and dreaming or had a vision, but suddenly the roof caved in. A large thing fell on the desk in front of her. It looked like an elephant. She climbed up on the edge of the desk and on top of the carcass, sure enough, it was an elephant. Being a nurse, she checked its eyes, etc, to make sure it was dead. It didn't have very big tusks. She thought to herself, "This thing doesn't have much value to it. It's going to start stinking pretty soon. If I have eat it, where do I start, the front quarter or a hind quarter? She was in 4H while growing up and finally decided that the hind quarter would have a little more fat on it. I can cut off the feet and make umbrella stands. Byrdeen was later relating the trance-like dream to her banker, Jeff Fagg. He started laughing and said, "That's exactly what you have on your hands, Byrdeen, a dead elephant."

There were a few close calls during the years of winding things up on the farm. One of the challenges was meeting the payroll each month. During the last year, just before harvest time, a $10,000 payroll was due within 72 hours and no money available. The bookkeeper, Petra Hoveland, was in tears. Byrdeen was pacing up and down the hall trying to stay calm. She was drinking coffee and saying, keep looking, see if we can't find a way to rob Peter to pay Paul. It was about that time Joe Guerro walked in and said, "I don't know what's wrong with my Dutchmen dairymen, but they really appreciate your hay. It's not even off the ground yet and they've already paid me, here's your share, $10,000."

A few months later the same situation came about. Byrdeen was in the office contemplating her lack of alternatives while probably sipping on the same cup, different coffee. The phone rang. It was the grain elevator calling. "You know, Don's had some grass seed here for quite awhile and the prices are really good right now. We had almost forgotten it, its worth about $10,000. Would you like to sell it?"

It took three years to sort out all the problems that come with a complicated estate. With her leadership and good people helping, they dismantled all the partnerships and corporations without the partners losing any money. She and the children retained five circles of farm ground.

Byrdeen worked as a nurse through much of the earlier years. She taught Sunday School at the Presbyterian Church, was a Cub Scout leader, active in the Republican Womens Club and serves on the Big Bend Community College Foundation Board. She has been active in the Black Sands Irrigation District. Her life in the Basin has been interesting and a good one, one thing for certain, said Byrdeen, "I've never been bored."

By the time Byrdeen got all the estate settled and things back to normal, she realized she was a stronger and more capable person. She could handle whatever life chose to hand her. It started with that oil drilling rig operator wanting to find a better way to make a living, to a dead elephant coming through the roof. She said life has been challenging, exciting and rewarding. We've had toys, trips, great children and family support and true friends. "We have been truely blessed." She felt very fortunate to ride the wave of those relationships in her times of need. When asked if she was ever sorry they located in the basin, Byrdeen said, "Never."

Brydeen

Don, Christine, Byrdeen and Scott

Don

Epilogue

The people who have allowed us to share their lives in the preceding pages are not unusual. They are typical of the settlers who built the Columbia Basin throughout the past 50 years with their determination to create a better life for themselves and their children. They had the same motives as the Puritans who landed at Plymouth Rock and the hardy adventurers that traveled the Oregon Trail.

Not even the farmer and rancher realize how important their industry is to the overall welfare of the general population. Yes, everyone knows you have to eat food to sustain the human body, but it is amazing how much two percent of the population effects the other 98 percent. It touches almost all segments of industry. Steel for trucks, tractors shipping barges, and trains all depend on farmers or their produce. Grocery stores, chemical and fertilizer companies, and nearly every business has some connection back to the agricultural base.

Recent studies have shown that each gross dollar coming from an irrigated acre creates another five or more dollars in off the farm related economic activity. Based on 1997-98 Washington Agricultural Statistics, Eastern Washington irrigated farms produced an average gross of $1,213 an acre. The Columbia Basin Project with over 600,000 acres contributes $4,366,000,000 to the economy in the northwest, and most of that in Washington State.

Add on to that figure all of the irrigated ground in Idaho, Oregon and California and it amounts to a staggering total. That is why the agricultural community needs to make the general public aware of the contribution farming has on the economy.

The air we breath is cleansed by growing crops using carbon dioxide out of the air. A corn field uses 18 ton of CO_2 per acre every year. Potato fields use 13 ton per acre. According to A.T. Brix, a chemical engineer from IPCI, an apple orchard uses 142 ton per year of CO_2.

We shouldn't be broadcasting how bad off farmers are, we should be telling them how important they are, not only because the U.S. spends the least percentage wise on the safest food available in the world, but also related job, environment and economic benefit to the country.

Index

A

Allan
 Clyde 148
Anderson
 Cody 21
 Dorothy 17
 Frank 17
 Martha 17
 Newel 20
Armstrong
 Larry 22

B

Bair
 Arlene 28
 Connie 15
 Dean 15, 17, 19, 20, 23, 24, 26, 28
 Dorothy 17, 20, 27, 28
 F. L. 15, 25
 Francis Leon 15
 Francis Leon, Jr. 15, 19, 25
 Glen 28
 Grace 15
 Jay 28
 Keith 21, 23, 28
 Laura Ethel 15
 Leigh 15
 Lyle 24, 28
 Myron 15

Ross 19, 28
Wayne 15
Baird
  John 136, 184
Bauer
  Al 88
Bell
  Arthur 210
  Bertha 210, 211
  Frank 210, 211, 212, 213, 214, 216, 218
  Frank, Jr. 210, 211, 212, 213
  George 211
Beltz
  Harold 52
Bergloff
  Dee 222
Berninger
  Jeff 107
  Joe 107
Bledsoe
  Stu 234
Bogart
  Becky
Boorman
  Cliff 171
Brix
  A.T. 252
Brown
  Aaron 136
  Juanita 167
  Marje 189
  Mike 189
Burke
  Vance 52

C

Cables
  Jim 101
Canfield
  Dave 28
Capone
  Al 222
Carlson

Index / 255

   Byrdeen 239
   Emil (Ted) 239
   Emily 239
Carson
   Kit 242
Cearns
   Dorothy 159
Chaplin
   Charlie 222
Charvet
   Andre 242
Chevalier
   Lt. Jacques 70
Child
   Ila 124
   Orville 135, 136
Christiansen
   Fran 226
   John 226
Cizik
   Irvin 225
Clapp
   Billy 21, 27, 218
Clay
   Bev 39
   John 38, 39
Clough
   Orville 74
Coleman
   Helen 238
Collins
   Cliff 52
Coolidge
   Calvin 222
Cooper
   Gary 222
Cox
   Dan 20
Curtis
   Catherine 147

D

Daniels

Archer 122
Davidson
  Fred 74
Dietz
  Lloyd 38
  Opal 38
Dodge
  Boyton 25, 203
Driggs
  Percy 166
Duirr
  Ann 226

E

Emtman
  Ray 202
English
  Jack 23, 224
Erickson
  May 42
  Wilford 42
Ertel
  Betty 35
  Ida 35
  Jack 35
  Janie 35
  Virginia 35
  Walt 35
  Walter 35

F

Fagg
  Jeff 245, 247
Field
  Addie 34
  Arthur 34
  Eddie 36, 37, 41, 43
  Janie 37, 38, 39, 42
  Kelly 36, 41, 43
  Marsha 36, 37, 41
  Marshall 35, 36, 39, 41, 42
  Robbie 39, 40, 41

Flanagan
    Sid 100
Flint
    Cathy 64
Ford
    Henry 222
Fullerton
    Gordon 134
    Penny 202

G

Gilcrist
    Bob 135, 136
Goodwin
    Bob 86
Gossett
    Cleo 51, 52, 54
    Dave 51, 52, 55
    Diane 51, 52, 55
    LeRoy 47, 51, 53, 54
    Renee 55
    Sharon 55
    Wayne 51, 54
Goulet
    Ken 100
    Roland 100
Graham
    Bob 63, 64
    Virginia 61, 63, 64
Greenwalt
    Arnold 88, 246
Grigg
    Lorin 122
Guerro
    Joe 247

H

Hagemeister
    Emily 239
Hagerty
    Claudia Lynn 73, 76
    Dean 67, 69, 70, 72, 75, 76

    Marie Clarice 75, 76
    Robert Wayne 74, 76
    Russel Dean 73, 76
    Russella 72, 75, 76
    Yvonne Louise 75, 76
Hale
  Mike 224
Hammond
  Bob 134
Hansen
  Frank 81, 83, 84, 86, 88, 89
  Ida Agatha 81
  Jerry 89
  Kay 89
  Penny 89
  Russ 83, 85
  Thomas Alvin 81
  Tom 89
  Wanda 84, 87, 88, 89, 90
Harding
  Warren G. 221
Harper
  Jay 120
Hart
  William S. 221
Henson
  Bud 225
Hess
  Bill 202
Hill
  Wanda 84
Hirai
  Betty 97
  Betty Marianne 97
  Haidi 93, 95, 96, 97, 119
  Jim 97, 228
  Kenny 97
  Nancy 97
  Patty 97
  Paul 101, 102, 105, 106, 108, 109, 122
  Shegeko 119
  Shigeko 95, 96, 97
  Shirley 97

Tom 97, 98
  Virginia 101, 102, 105, 107, 109
Holloway
  Agnes 121, 205
  Betty 95, 115, 119
  Bob 86, 88
  Dale 124
  Earl 64, 95, 113, 114, 118, 119, 121, 122, 123, 125, 169, 171
  Frank 115
  Marion 116
  Robert 115, 121, 122, 169, 205
  Ruby 95, 113, 114, 118, 119, 121, 122, 123, 125, 171, 232
  Thelma 123
Holman
  Paul 86
Hovland
  Petra 246, 247
Hull
  Dianne 132, 137
  Judy 132, 137
  Marianne 137
  Merle 131, 132, 136
  Richard 132, 137
  Robert 132
  Ron 137
  Roy 23, 131, 132, 135, 136, 162
  Steven 132, 137
Hussey
  Shirley 83
Hyer
  Chris 134

I

Iseri
  Shigeko 94

J

Jenks
  Parley 118
Johnson
  Cleo 47, 49
  Eric 28

Jolson
  Al 214, 222
Jones
  Sellden 71

K

Kaynor
  Gib 74
Kelly
  Percy 135, 136
Kniep
  Elmer 43

L

Lange
  Willard 26
Leach
  Roy 122
Lee
  Patsy 104
Lobe
  Everett 120
Lauzier
  Paul 232

M

Marcuson
  Duane 136
Masto
  Harry 102
Mathews
  Verne 27
Matthews
  Gail 218
Mauk
  Eleanor 196
May
  Catherine 86
McAfee
  Charlie 36, 37
McCarroll
  Tom 16

Index / 261

McMullen
  Harry 27
Mix
  Tom 222
Moore
  Carolyn 140, 147
  Catherine 147, 148, 150
  Dean 140, 141, 143, 144, 145, 146, 148, 150
Morris
  John 155, 156, 171
  Johnny 122
  Paul 171
  Shirley 171
Murphy
  Andy 164
  Bill 164
  Bob 164
  Dorothy 159, 163
  Jerry 164
  Jim 164
  John 164
  Ken 135, 159, 160, 161, 163, 164
  Kevin 164
  Larry 164
  Mark 164
  Maureen 164
  Mike 164
  Rebecca 164
  Teresa 164
Myers
  Irwin 43

N

Nalder
  Phil 163
Nasburg
  Ed 21
Nave
  Bessie 118
  Ted 118
Neal
  Ralph 172
Newton

Zane 43
Nomoto
  Yoshio 108

O

Ottmar
  Brenda 168, 172
  Harley 165, 166, 168, 169, 170, 171, 172, 173
  Irene 166
  Juanita 167, 169, 173
  Rich 170
  Vic 166, 168, 170, 172, 173

P

Parker
  A.E. 177, 179, 182
  Alice 180, 181, 182, 183, 186, 188, 189
  Carla 184
  Hazel 179, 182
  Ike 177, 179, 181, 182, 183, 185, 186, 188, 189
  Perry 184, 187
  Susie 184
Patton
  Russ 27
Pearl
  Dean 122
  Ed 122
  Larry 122
Petrak
  Bill 134

R

Rainwater
  Jacob S. 73
Raugust
  Bill 167
Rayburn
  Glen 74
Reese
  Darrell 245, 246
Richardson
  Annette 198

Bob 198, 201
  Carole 198
  Doug 198, 203
  Eleanor 196, 197, 198, 200, 201, 202, 204, 205, 206
  Greg 203
  Larry 196, 197, 198, 200, 201, 203, 204, 205
  Larry, III 203
  Larry, Jr. 203
  Laura 200, 203
  Teresa 203
Ritz
  Georgia 104
  Ron 104
Romano
  Jane 205
  Tony 170
Roumpf
  Larry 225
Roumph
  Larry 40

S

Saunders
  Bud 86
Schwab
  Harold 86
Scott
  George 26
  Stuart 64
Seivers
  Gert 39
Sheehey
  Leo 135, 162
Shulz
  Gene 224
Snead
  Mode 74, 75
  Tim 186
Stetner
  Rudy 203, 204
Stewart
  Jimmy 222

T

Taylor
  Ralph 180
  Sara 180
Thompson
  Ben 210, 214, 215, 216, 218
  Bertha 210, 218, 219
  Mabel 210, 211, 212, 213, 214, 215, 216, 218, 219
  Myrtle 209
  Sarah 218
  Winifred Russella, 68
Thornton
  Dee 149, 222, 225, 226, 229
  Everett 149, 221, 223, 225, 226, 228, 229
  Jack 224
  Jane 224
  Linda 223, 224
  Mark 224, 229
Tobin
  Jack 40
Townsend
  John 104, 107
  Mike 104
  Vern 118

V

VanHoltz
  Adrian 170

W

Washington
  Nat, Sr. 218
Watkins
  Esther 43
Weber
  Jeanne 40
Wetzel
  Everett 84
Williamson
  Alan 233, 235
  Charlie 231, 234

Diana 235
  Ed 231, 233, 235
  Jerry 232, 235
  Kathy 235
  Larry 232, 235
  Marcia 234, 235
  Melody 235
  Signe 231, 233, 235
Wiser
  Emery 134
  Wayne 134
Worley
  Byrdeen 238, 239, 240, 241, 242, 243, 244, 245, 247, 248
  Don 238, 239, 241, 242, 243, 244, 245, 247
  Harvey 238
  Helen Louise 238
  Kristine Dianne 243
  Scott Harvey 243

Y

Yada
  LaDel 104
Yorgenson
  Frans 20
Yoshino
  George 170

to order additional copies of

Desert of Dreams

call

(800) 917-BOOK (2665)

email

agbob@bossig.com

visit

www.selahbooks.com